SEXLESS OYSTERS AND SELF-TIPPING HATS

BY ADAM WOOG

SASQUATCH BOOKS

SEATTLE, WASHINGTON

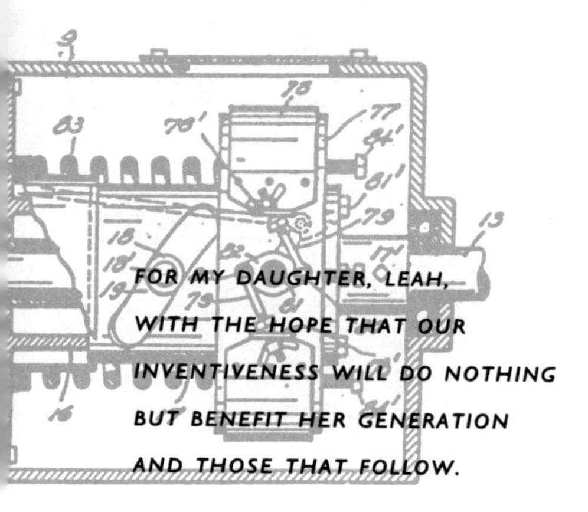

FOR MY DAUGHTER, LEAH, WITH THE HOPE THAT OUR INVENTIVENESS WILL DO NOTHING BUT BENEFIT HER GENERATION AND THOSE THAT FOLLOW.

Library of Congress Cataloging-in-Publication Data
Woog, Adam, 1953–
Sexless oysters and self-tipping hats: 100 years of inventions in the Pacific Northwest / Adam Woog.
p. cm.
Includes index.
Summary: Surveys 100 years of inventions in the Pacific Northwest, in such fields as outdoor gear, games, fishing and shipbuilding, and medicine.
ISBN 0-912365-47-1 : $14.95
1. Inventions—Northwest, Pacific—History—Juvenile literature. [1. Inventions—Northwest, Pacific—History.] I. Title.
T30.A1W66 1991
609.795 — dc20 91-24522
CIP
AC

Designed by Scott Hudson, Marquand Books, Inc.
Typesetting by Scribe Typography.
Printed in the United States of America.

Published by Sasquatch Books
1931 Second Avenue
Seattle, WA 98101
(206) 441-5555

Contents

Preface

It shames me to say that I was a little skeptical when I first began to explore the world of Pacific Northwest inventors. Loyal as I am to my native ground, the Northwest didn't spring immediately to mind as a hotbed of American invention. Surely all the good stuff happened — is happening — in the Midwest and Northeast, with their longer histories of industrialization and denser concentrations of brain-power?

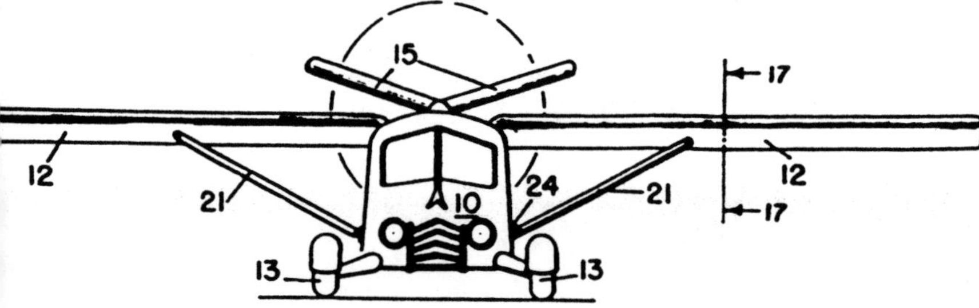

Would there be enough material in the Northwest to warrant a book? Were that many interesting things really invented here?

As soon as I dug into the subject, however, I unearthed an overwhelming amount of stuff — so much, in fact, that I had to limit the territory to Washington and Oregon. Even then there was too much, and I had to decide who and what to leave out. The result is by no means comprehensive, but I hope that it is representative.

In some cases, entire areas of endeavor have been given short shrift. Biotechnology and genetic engineering, for example, are subjects of growing importance in the Northwest, but the field is so new that many companies are still years away from having products on the market. I settled for letting the triploid oyster represent the entire topic. In other cases, I stretched the definition of "invention" a bit; if I had limited myself to bona fide patented items only, I would have had to leave out a lot of good stories. In still other cases, the true origins of an invention have been lost, or the legends are vague, or the stories conflict. I tried to present the truth as much as possible, of course, but as G.K. Chesterton

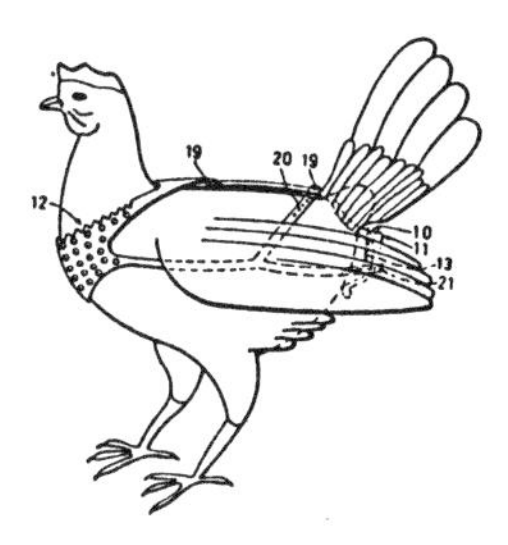

said: "A touch of fiction is almost always essential to the real conveying of fact."

Speaking of fiction, many of the suggestions I pursued led to dead ends. One informant swore up and down that Velcro had been invented in the Pacific Northwest. (A Swiss textiles expert named Georges de Mestral created it in 1948.) Due to a miscommunication, I labored for a brief but excited time under the illusion that Monopoly had been invented in Oregon. (It was developed in 1933 in Germantown, Pennsylvania.) I was told that a former chemistry teacher at the Lakeside School in Seattle invented Fizzies. (He patented a means of isolating a particular chemical that had nothing to do with soft drink fads.) And there are some who will tell you that the Pet Rock came from Washington State. (California, actually.)

As I worked, I tried to find some common thread that linked my subjects. Are Northwest inventors different from other inventors? On the whole, I must admit that Northwest inventors share the attributes of inventors anywhere: stubborness, individualism, creativity, perseverance, industriousness, intelligence. The spirit that drives them, I think, is what energizes creators anywhere: the urge to tinker, to improve, to make things better or faster or richer or deeper. Where inventors live is usually a matter of happenstance.

But the same is not true for their products. A spirit of inimitable Northwestiness is found in these pages. Many inventions can be traced directly to the Pacific Northwest climate and geography: outdoor gear makes a strong showing, as do machines and techniques for logging, agriculture, and fishing. It makes a certain amount of sense that the water ski and the hydroplane were developed on the shores of Lake Washington, and that Oregon's timber country inspired loggers to harvest trees with dirigibles and helicopters. What better place to claim the origins of a superefficient wood stove? On the other hand, there are plenty of inventions that were born here strictly by chance — how else to explain the Northwest origins of the self-cleaning house, the electric guitar, the heart defibrillator, or even the Space Needle?

Since only a sampling of the inventions I unearthed could fit in one book, I still have a fat file of "possibilities." There are hundreds or even thousands more out there about which I am still ignorant. I hope that readers where are aware of other Northwest inventions will contact me c/o Sasquatch Books, 1931 2nd Ave., Seattle, WA 98101. I would also be grateful to anyone who can clarify uncertainties or correct mistakes.

Foreword

I have never understood the inventor's mind. The closest I've come to innovation is in a consistent failure to follow printed directions. This has not been productive. To the best of my knowledge I've never met an inventor, so when Adam Woog mentioned to me that he was gathering material for a book on regional inventions, I feared he would find the pickings slim. How delightfully wrong I was.

It turns out that our Northwest woods are full of inventors — our seashores too, even Eastern Washington. True, some of them are kin to the inventor portrayed by Jonathan Swift in *Gulliver's Travels*, a dreamer who spent "eight years upon a project for extracting sunbeams out of cucumbers, which were to be put in phials hermetically sealed, and let out to warm the air in raw, inclement Summers." Northwest offerings to the unappreciative world have included the dog toothbrush, the self-tipping hat, the city-circling mechanical beltway, and the Tri-Phibian, a fluorocarbon-powered land-sea-air vehicle.

But don't be misled. This is not a catalog of nuts. The nuts garnish the solid fare of local inventions that have influenced our lives, sometimes for the better: Elmer's Glue, Boeing's 727, Dvorak's keyboard, Pocock's rowing shell, and Edmark's heart defibrillator.

The inventors, briefly profiled in all their diversity, range from shopkeeper-sportsmen like Eddie Bauer to immigrant visionaries like Constantinos Vlachos, from the professional bike-rider and rowing coach Hiram Conibear to the pseudonymous Frances Gabe, perfecter of the self-cleaning house.

Adam Woog is a canny researcher with a gift for explaining the principles involved in such things as air-cushion vehicles, sonic holography, and the oscilloscope. He possesses a mind unmindful of boundaries. How else could he stretch the concept of invention to include D. B. Cooper's skyjack and Kenneth Arnold's flying saucer?

Sexless Oysters and Self-Tipping Hats is crammed with information, both useful and inconsequential. It offers a fresh perspective on the economic history of Washington and Oregon. And it is great fun.

Murray Morgan
Trout Lake

Inventions for the Great Outdoors

Sno-Seal Boot Wax, Penguin Sleeping Bag

1930s

Ome Daiber is a legendary figure among Northwest mountaineers: a celebrated outdoorsman in a region filled with expert climbers and devoted hikers. During the 1930s and 1940s he helped establish dozens of new routes into the remoter parts of the Cascades and Olympics, and was also a key figure in the formalization of mountain rescue operations in the region — an idea he borrowed from European climbing circles and helped to adapt for the Mountaineers club of Seattle and other organizations. Daiber was also the inventor and manufacturer of a number of outdoors-related products.

He was born George Daiber in West Seattle. (He later legally changed his name to reflect his longtime nickname. The story goes that when Daiber asked a friend who ran the West Seattle High School cafeteria to front him lunch money one day, he said, "Lend me a quarter and you'll owe me," and the phrase stuck.) An avid hiker and climber, for years Daiber operated the Seattle Boy Scout trading post and an outdoors shop called the Hike Shack. In the 1930s, frustrated at the lack of inexpensive, high-quality outdoor gear available, he formed Ome Daiber Inc. to manufacture and retail tents, garments, quilts, and other pieces of outdoor gear, many of which were of his own design. But the company became too big and stressful for his liking, and he found himself with too little time to spend in the mountains. In 1943, he sold the manufacturing rights to many of his products and became a carpenter, builder, and remodeler of houses and commercial buildings.

The best-known of Daiber's creations is Sno-Seal, a

Ome Daiber's most successful invention was Sno-Seal (right), but he had some interesting failures as well, such as the Zipelope, an envelope that opened when a string was pulled, Band-Aid–style.

protective coating for leather boots that was the standard water repellent in the Pacific Northwest for many years. It is still available, still packaged and sold in its original, distinctive blue-and-white can. Daiber developed it in the mid-1930s, his widow Matie says, because "he needed something to keep his feet dry that was better than Hubbard's Shoe Grease, which was about the only thing available then. He couldn't find the right stuff, and he didn't have much money anyway, so he couldn't afford even what he did find." Sno-Seal's main ingredient is beeswax, which is more water-repellent than previously used oils. Daiber found a secret way to combine it with various oils so that the solution penetrated boot leather without rubbing off.

In the 1930s, Daiber also invented the Penguin sleeping bag, "the only sleeping bag in the world offering the unique feature of permitting the user to get up without getting out of bed." (He filed for a patent on the Penguin in 1935, although it was not granted until 1939.) The Penguin was a goose-down sleeping bag with built-in arms and legs that

Daiber and his friend the Penguin on the Hoh River in 1934.

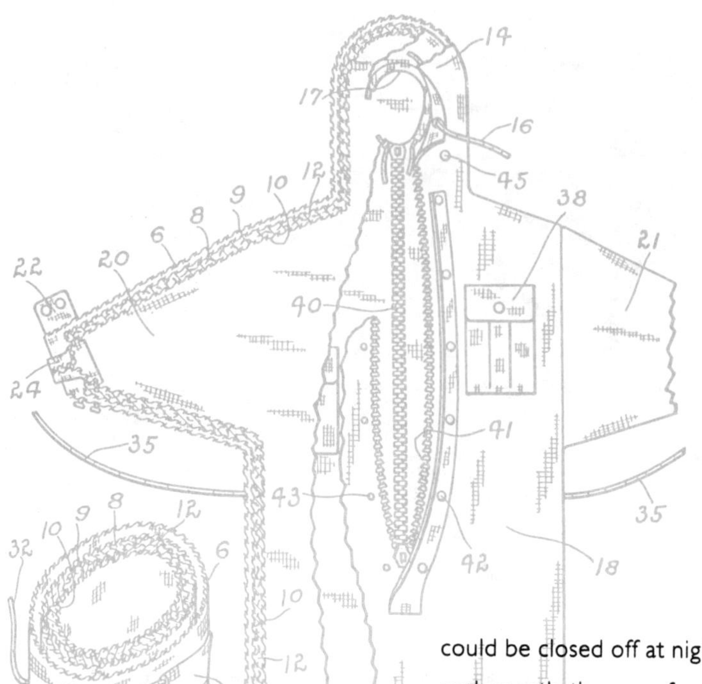

could be closed off at night with ties; the arms were tucked underneath the user for extra cushioning while the user's legs went into a single leg of the bag. The bag's hood came over the head and had a cloth extension, shaped like a long elephant's trunk, that funneled the user's breath outside and kept it from condensing inside the bag on very cold nights. The bag could be worn around camp in lieu of a parka; horsehide on the bottom of the "feet" kept the canvas from tearing. Only about 90 were manufactured in total. Matie Daiber says the bag was a commercial failure because of its expense: "No one who wanted one could actually afford one."

The Trapper Nelson Pack

1920

The Trapper Nelson (modeled here by its inventor and his daughter Lois) was a stalwart standby until the mid-1970s, when it was finally done in by high-tech packs of aluminum and nylon.

The Trapper Nelson packboard, the world's first mass-produced frame backpack, was a severe and rather Spartan item, used mainly by earnest Boy Scouts and hardworking firefighters. Its canvas sack was only feebly waterproof, the wooden frame was far from comfy, and — compared with other outdoor gear of the 1920s — it was expensive: $4.50 for the small model, $5.50 for the medium, $6.50 for the large. By today's high-tech standards it was a dinosaur, but its appearance marked the beginning of a new era. It helped begin a revolution in hiking gear that fit perfectly with the Northwest's burgeoning love for the outdoors. The pack inspired fierce loyalty long after it was overtaken by superior technology. *The Seattle Times'* John Reddin remarked in 1966 that, "like family heirloom silver, the rugged Trapper Nelson apparently goes on forever, the time-proven favorite of Boy Scouts, forest rangers, hikers, hunters, prospectors and others who travel by foot in the great outdoors."

The pack was born in 1920, when Lloyd Nelson, a civilian employee of the Puget Sound Naval Shipyard in Bremerton, was asked to check shortages at a construction site on Wood Island, near Kodiak. "When my work was finished," Nelson recalled for *The Seattle Times* in 1961, "I asked for and received a short leave without pay, and joined the throngs of miners, fishermen and other foot-loose Alaskans who were staking claims on newly open oil-reserve land. While assembling an outfit to enable me to cross a mountain range on foot, an Indian agreed to lend me his crude Indian packboard made of sealskins stretched over willow sticks, a style used for generations of his ancestors. I made the trip and staked my claim, but afterwards lay awake nights recalling the back-breaking ordeal and wondering if it would be possible to evolve a really comfortable backpacking device."

At home in Bremerton, Nelson set to work on a version of his anonymous benefactor's pack. Anything would have been an improvement on the crude rucksacks then available, which were little more than bags with shoulder straps. Nothing kept the contents from gouging the wearer's back, most of the weight was borne by the hips, and there was no way to keep from bending forward under a heavy load. Nelson designed a frame that distributed weight more evenly on the shoulders, a canvas jacket that cushioned and ventilated the back, a removable sack to make loading easier, an outside pocket to hold small objects, and a large top flap to accommodate a bedroll. He then invested in a sewing machine, a bolt of canvas, thread, grommets, dies, cord, buckles, a side of leather, and a cutter. He hired a young Bremerton man named J. D. ("Dorm") Braman, later a Seattle councilman and mayor, to make experimental frames of spruce and oak. Although not a dedicated outdoorsman — that hike in Alaska was apparently his one big fling — Nelson tested his prototypes on numerous camping trips, and then arranged a manufacturing production schedule with the Trager Company, a Seattle maker of logging gloves and aprons.

For most of the 1920s, Nelson did his best to sell the pack on his own, sending free samples to prominent outdoorsmen and making frequent visits to sporting-goods stores. (He often walked into shops carrying his young son in the pack to demonstrate its durability.) But these were the days before the great outdoors was chic; the rough-and-ready sportsmen who spent serious time in the woods weren't interested in fancy items. The consensus among store owners, Nelson remarked later, was that "my product was too good-looking for the type of person who carries food, clothes, and blankets from place to place. The best I could do was leave my samples on consignment."

After years of only marginal success, Nelson finally sold the manufacturing rights in 1929 to Charles Trager for $5,000. In retrospect, it was an unfortunate move for the inventor. Two weeks after the deal was signed, a rush order came from the U.S. Forest Service in Missoula; 500 packs were desperately needed for their firefighters. Two weeks later came another order for 500, this time from the Forest Service office in Salem. Requests soon began pouring in as the U.S. Geodetic Survey, the Boy Scouts, and numerous other groups — including the Abyssinian Water Development Commission, which ordered 50 — began to realize its usefulness. The Trapper Nelson had finally arrived.

Down Parka

1934–36

Eddie Bauer shows off the down vest that made him famous.

Seattle outdoorsman and entrepreneur Eddie Bauer, along with such people as Maine's L. L. Bean and the group of hikers who started Recreational Equipment Inc. in a West Seattle living room, helped change the way the American public thinks about the wilderness. By forging an association between high-quality camping gear and a high quality of life, they virtually created the notion that expeditions to the Great Outdoors need not be the exclusive province of crusty eccentrics.

Bauer's most significant innovation was the quilted down parka, which he patented in 1936. It was a genuine landmark in both the sports industry and the clothing world, revolutionizing outdoor wear and helping spawn the style known as Outdoor Chic. Although Bauer himself did little to foster the trend toward stylishness, much less to promote the specific fashion Tom Wolfe once called "the hand-grenade look," his down jacket made Bauer rich and famous.

Born on Orcas Island in 1900 to Russian immigrant parents, the young Bauer was happiest when fishing, hunting, or riding horseback. When the family moved to Seattle in 1913, he compensated for the loss of access to the wilderness he loved by clerking at Piper and Taft, then Seattle's largest sporting goods store. He was a star salesman by his mid-teens, but began to concentrate on his special gift: stringing tennis rackets. He won a world championship by stringing a dozen tournament-quality rackets in 18 minutes apiece. He used a machine of his own invention, a foot-powered tool for maintaining string tension, which he patented but was

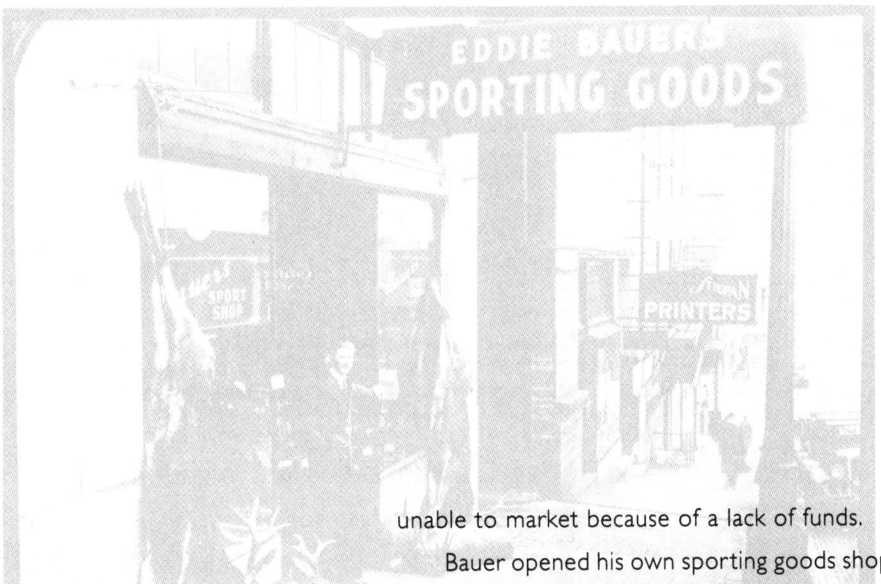

unable to market because of a lack of funds.

Bauer opened his own sporting goods shop in 1922. His customers knew that the store's owner personally put every piece of equipment, from fishing rod to sleeping bag, to the test. If Bauer himself wasn't satisfied with something, he created a new and improved version. This hands-on policy led directly to the development of the down jacket. While fishing the Humptulips River on Washington's Olympic Peninsula during the winter of 1934, Bauer nearly froze to death. He had discarded a too-heavy woolen jacket and come dangerously close to contracting hypothermia. Thinking later about ways to make outdoor clothing both lighter and warmer than the traditional woolen jacket, he remembered a Russian uncle who had always claimed that quilted goose-down undergarments kept him alive during the 1904 Russo-Japanese War. As it happened, Bauer was already importing goose-down feathers from China, to use for fly-tying and for the manufacture of an improved badminton shuttlecock he had invented. He made a couple of quilted down jackets for himself and friends, and they were such hits that he patented both the design and the process.

The first jackets Bauer manufactured were made of

canvas, with the material sewn into double-thick, diamond-shaped pockets with one edge left open. Bauer designed a machine that measured the amount of down needed for each quilted pocket; this amount varied, depending on the ultimate use of the garment. A jacket designed for extremely cold weather, for instance, would use more down in the arms than in the torso to compensate for the generally poorer circulation in the extremities. The blowing machine (which John Kime, a longtime Bauer staff member, says "looked like a ten-gallon can attached to the business end of vacuum cleaner") blew the down into one pocket at a time; the pocket had to be stitched shut and the machine refilled before the process could be repeated.

The "Skyliner" — the world's first commercial down jacket, and still one of the most popular items in the Eddie Bauer catalog — went on the market in 1936; the "Canadian Downlight" quilted vest came out the following year, and both products met with instant success. Kime says the company couldn't keep up with demand, and back orders of six months were common.

World War II was both a curse and a blessing for Bauer's business. The U.S. government requisitioned all available down supplies, though Bauer was allowed to make garments for cold-storage workers and doctors working in severe climates. But the military also ordered a quarter of a million sleeping bags and thousands of specially designed high-altitude bags for wounded soldiers (with zippers all around so that any part of the body could be reached for medical treatment), as well as 25,000 down flight suits — trousers with suspenders and hooded jackets. Bauer insisted that each of his products carry his logo; they were the only military-contract items to do so.

Quick-Release Ski Binding

1937

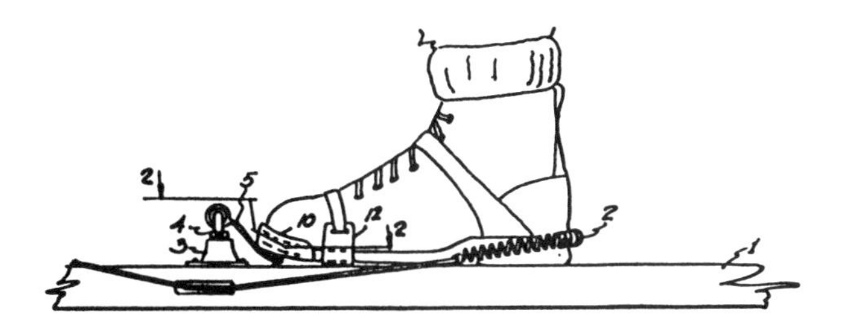

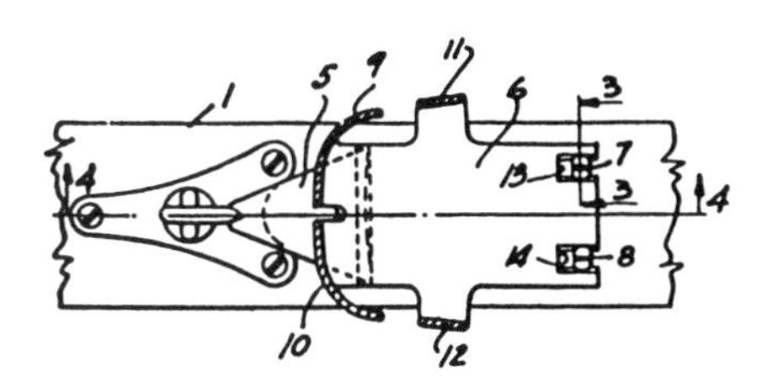

Besides inventing a device that has kept thousands of legs from breaking, Hjalmar Hvam was an outstanding competitor in the now-defunct four-way class of skiing (downhill, slalom, cross-country, jumping). He was elected to the National Ski Hall of Fame in 1971.

In the spring of 1937, after he had successfully defended his status as champion in the Golden Rose Race on Mount Hood, Hjalmar Hvam, an immigrant from Kongsberg, Norway, went free-skiing with friends. Hvam jumped a crevasse and twisted to avoid an icy piece of cornice. He fractured his leg while essentially standing up, and was taken to Portland to have his leg set. After the operation, he was struck by an idea about how a skier might avoid such an accident in the future. "Just as I came out of the ether, I had this idea. I called for the nurse to bring me a pencil and paper right then and there. When my wife came to the hospital the next morning, I said, 'Now I have the safety binding.' "

A mechanical draftsman by training, Hvam had just invented the world's first quick-release ski binding. Previously, bindings were crude and dangerous affairs known as "bear traps." They consisted of fixed heel-spring and metal toe clamps, with an adjustable leather strap or metal cable that wound around the ankle and toepiece to keep the boot snug. Where the ski went, so went the skier, and in the case of a fall there was no way to jump clear.

Hvam's improvement was a pivoting metal toepiece, shaped like a Y, that held the toe of the boot firmly in place

under normal conditions. A spring kept it aligned properly as long as there was forward pressure on it from the ski-boot toe. At the first sign of any forceful movement to the right or left, however, the spring buckled and automatically threw the foot out of the binding. The heel remained strapped to the ski so that the ski would not become lost, but the skier's ankle was left free.

Hvam patented the binding in 1939, and it was an immediate success. He quit his job operating the rental shop at Mount Hood's Government Camp, set up his own ski shop in Portland, and formed a company to manufacture the binding (his sales slogan was "Hvoom with Hvam"). The model outsold the rest of the market for a decade, averaging 10,000 to 15,000 units annually. It remained popular into the 1960s, and *Ski* magazine notes that "some still think [it] is the simplest, most elegant release binding ever made." An example is on display at the Oregon Historical Society Museum.

Water Skis

1928

Right: Don Ibsen ditches one of his water skis. Below: The latest in lake transport.

One day during the summer of 1928, Don Ibsen, a recent graduate of Seattle's Roosevelt High and an avid sportsman, had a bright idea: if he could ski on snow, why not on water?

Ibsen began his experiments on Lake Washington by making a crude pair of aquaplanes, then called surfboards, out of wooden boxes. Tying a rope to the back of a friend's speedboat, he simply stood on the aquaplanes (without attaching them to his feet) and grabbed on. They worked, but not too well. Next he tried regular snow skis, but these were too narrow; he lashed a pair together, but the water kept shooting up in between. Finally, he carved two slabs of cedar, each about eight inches wide and seven feet long, and hand-bent them by steaming them over a five-gallon can of hot water and staking them against a telephone pole. He sanded and shellacked them, and nailed on an old pair of tennis shoes. These skis provided the necessary buoyancy without bogging down the rider, and that summer Ibsen became a familiar sight on Madrona and Leschi beaches, sitting around with his odd-looking skis and bumming rides from passing motorboats.

That winter, Ibsen went to work as a hardware salesman. During the late 1920s and early 1930s, he made water skis for himself and friends, but soon was making them for sale. His first pair sold for $19.95 in 1934. (Some of these early skis are on display in Seattle's Museum of History and Industry.) But several factors — a Depression-era lack of funds and a wartime scarcity of materials, plus an attorney

who was called up in the middle of negotiations — combined to keep him from patenting his new creation. "I sent the dope about it [to the U.S. Patent Office] back East," he recalls, "but nothing much ever happened. I didn't have any money to pursue it, anyway."

The lack of a patent didn't bother Ibsen. Echoing a familiar sentiment among inventor-entrepreneurs, he says, "I've always believed the best patent is being there first with the most." From all accounts, Ibsen in his prime epitomized the classic selling-ice-cubes-to-Eskimos variety of pitchmen ("They told me, 'Ibsen, with your mouth you'd better become a salesman' "). He was so successful at marketing his new product that he quit the hardware business and devoted himself full-time to marketing Ibsen Water Skis and related equipment, later expanding into the promotion of "water fun" shows and water-skiing races around the country.

Ibsen's many exploits include water-skiing to work across Lake Washington, complete with briefcase and suit, a stunt photographed for *Life* magazine. He also took the part of the groom during a mock water-ski wedding for a San Francisco promotional stunt, during which all the participants were towed around San Francisco Bay by helicopter. (Ibsen's son, playing the priest, ended the ceremony by intoning, "I now pronounce you wet.") Although he claims to have been "a real chicken" when it came to dangerous tricks, Ibsen has an artificial knee as a result of his water-skiing exploits, and acknowledges that he would water-ski on occasion with a girl

"Dud" Davidson in a free-fall jump, 1944.

on his shoulders "if she was pretty enough."

Ibsen was also towed by helicopter across Lake Washington as part of the opening ceremonies for the Evergreen Point Floating Bridge. He has skied behind seaplanes, automobiles, motorcycles, and even a horse. East of the Cascade mountains on a sales trip one summer before the war, Ibsen spotted a young woman riding a horse beside an irrigation ditch: "I thought, 'Hell, that's the way to do it.' " He happened to have a 75-foot towline in the trunk of his car, so he caught up with the woman and convinced her to pull him on his water skis behind her as she galloped along the ditch. "It was quite a surprise to her, but it worked fine. So now when these young guys brag about how many horsepowers they're towed by, or complain about not having enough horsepower, I just tell 'em I was once towed by a *single* horsepower."

Hydroplane

1929–50

In the first years of this century, racing boats were skinny and long, often compared to watergoing cigars. At least one extra person, usually a mechanic, rode with the pilot to take care of the inevitable breakdowns. The boats had ordinary "displacement" hulls — that is, they plowed through the water, splitting it with their V-shaped hulls and displacing a considerable amount of it with their volume. The concept of using aerodynamic principles to raise the boat from the water, thus reducing the amount of friction between the water and the hull, was first tried in the 1920s by a number of different designers, but the idea was perfected and made practical by a team of three Seattleites: Anchor Jensen, a second-generation boat builder; Stan Sayres, a wealthy car dealer who was also a raceboat sponsor and owner; and Ted Jones, an engineer and boat designer. The boats they assembled together, *Slo-Mo-Shun III* and *Slo-Mo-Shun IV*, became paradigms for the world of unlimited hydroplanes. They helped bring the Gold Cup to Seattle and to make the annual Seafair races on Lake Washington an enduring summertime ritual.

By 1957, the Gold Cup class (which prohibited jet propulsion and movable airfoils) had essentially disappeared; the powerful and faster "unlimited" boats proved more popular with racing fans. Current regulations for unlimited boats state that they may be 28 to 40 feet long; have any type or number of inboard engines, though no jet propulsion is permissible; must be steered by a submerged rudder, with no aerodynamic devices; and must weigh over 5,500 pounds without fuel.

The hydroplane's characteristic "roostertail" plume occurs because the propeller, a third of the boat's "three-point" design, is only half in the water.

The question of who really perfected the three-point hull design — hallmark of the modern hydroplane — is still open to debate. Undoubtedly, Jones and Jensen both contributed greatly to the boat's high performance, although Jones, a genius at self-promotion, claimed the lion's share of responsibility. For his part, Sayres served as both a catalyst in bringing Jensen and Jones together, and as the financier who made their expensive experimentation possible.

After graduating from Seattle's Broadway High, Jones was an auto mechanic, an amateur boat builder, and an airplane enthusiast before he joined Boeing as an engineer in 1937. (His first boat, according to legend, was built from the plans that came with a 45-cent can of glue.) Lacking the money to make his homemade boats go faster by simply increasing their engine size, he turned his attention to redesigning their hulls for greater efficiency. As early as 1929, he experimented with bulges (called sponsons) that angled at 45 degrees from each side of a speedboat's hull. These provided greater stability and enabled him to put the boat through high-speed turns without flipping. By 1934, he had built a prototype three-point hydroplane that incorporated sponsons and two outriggers forward. It performed poorly on the turns and tended to slip sideways in the straightaways, but it was, he recalled later, "just fast as hell."

Jones's hydroplane was still a highly unpredictable prototype when he formed a partnership with Sayres in the late 1940s and early 1950s. Sayres asked Jensen, a boat builder and repairman on Lake Union, to build *Slo-Mo-Shun III* from Jones's design. Jensen not only built it, but also significantly reconfigured it after Jones was stymied by design elements during its construction. *Slo-Mo-Shun III* was radical for its time, wild and unpredictable. The project did not become truly successful until the trio built *Slo-Mo-Shun IV* after a trip to Detroit to check out the competition at the 1948 Gold Cup races.

The historic trial run on Lake Washington, October 1949.

Jensen recalls that "it was Sayres's intention to build an unlimited hydroplane...it was his dream to put something together that would go faster than anything else. At the time...they were only running displacement boats [in Detroit], with the propeller in the middle and a single-stepped hull. They were well done, but not effective. They were too heavy and awkward. You need something that rides right on top of the waves, strong and light. All the boats were tail draggers — in other words, they weren't getting up on top and making a roostertail. They did have one boat that looked good — the *Hurricane*, built by Grumman. They'd put lots of money into it, and it had lots of streamlining, but she didn't ride right. She was still dragging her rear end. And streamlining is detrimental to a hydroplane — they had aircraft engineers working on it, and streamlining is good for planes, but with a boat the potential for the main lift is from the water. You don't want air lift."

The trio returned to Seattle convinced they could build a better racing boat than anything they had seen. *Slo-Mo-Shun IV* used Jones's basic three-point design but incorporated a number of Jensen-inspired modifications, including the use of a single rudder instead of two, a sharply defined, knifelike nose (with no rounded surfaces that could slow the boat down), and a gearbox with a 3-to-1 ratio instead of the previously used 2-to-1 ratio. The engine they tried first was an Allison that Sayres had bought as surplus, but it was later replaced by a British-built Rolls-Royce Merlin. With the Merlin installed, Jensen attacked and solved a crucial problem: how to reverse the position of the engine air intake, from the bottom (where it was normally installed on an aircraft to provide visibility for the pilot) to the top. The reversal was required so that the engine's carburetors would not be positioned in the boat's bilge, where they might backfire and

Left: On October 1, 1949, Anchor Jensen and Ted Jones are towed in *Slo-Mo-Shun IV* in preparation for its maiden voyage on Lake Washington. Note the four-masted schooner in the background. Above: Seafair summertime blowout, 1983.

Photo ©1991 Mary Randlett

send the boat up in flames.

The new hydroplane became known as "the Hunt's Point ferry" because it made so many crossings of Lake Washington during its two years of development. But the testing paid off: in 1950, *Slo-Mo-Shun IV* set a stunning world straightaway record of 165.3225 miles per hour at the Gold Cup in Detroit and shattered the previous record of 70.89 miles per hour, set by Danny Foster in 1947 in the *Miss Peps V*. That year, *Slo-Mo-Shun* not only took the victor's place in the Gold Cup races, establishing a winning streak that lasted through 1954, but brought the races to Seattle and permanently placed Seattle's summertime Seafair celebrations at the center of competition hydroplane racing.

After their initial Gold Cup victories, Jones designed and built another boat for Sayres, *Slo-Mo-Shun V*, using Jensen's boatyard as well as Jensen himself; but the working relationship between Jensen, Jones, and Sayres had always been tenuous, and Jones left the partnership to work for the president of Mercury outboards. He later produced a number of other champion hydroplanes, including the *Misses Bardahl*, the *Misses Thriftway*, and *Rebel Suh*. Sayres continued to sponsor hydroplanes until his death in 1956, Jensen still builds boats in the Lake Union shop his father built, and *Slo-Mo-Shun IV*, the single most important benchmark in the history of the modern hydroplane, is now owned by Seattle's Museum of History and Industry.

Waffle Sole for Running Shoes

1971

The modern "athlete's shoe" appeared soon after Charles Goodyear invented vulcanized rubber in the 1860s, and by 1894 the Spalding Company was selling spiked shoes for six dollars. Athletic footgear has been called various names over the years — sneakers, gym shoes, tennis shoes, runners, and plimsolls (so named because British sailors wore them equipped with a mud-guard that resembled the Plimsoll mark used to denote legal water levels on ships). Keds, the first widely marketed modern sneakers, were introduced in 1917 and had tan canvas uppers and black rubber soles embossed with a shallow, tirelike pattern. There were also more sophisticated spiked shoes for serious track-and-field athletes, which pioneered the use of lightweight materials. But for the general public, the basic Keds-style tennis shoe (called "flats" by serious runners and used by them mainly for warming up) remained virtually unchanged until the early 1960s, when two Oregon athlete-entrepreneurs, Bill Bowerman and Phil Knight, started a revolution.

Even without his contribution to running shoes, Bill Bowerman would be a legendary figure in the sport of track. As track coach at the University of Oregon in Eugene, which for years has been a major center for running, he coached NCAA championship teams in track and cross-country. (He coached the 1972 U.S. Olympic team as well.) His 1966 pamphlet "Jogging" played an important part in laying the foundation for the current running boom, and his theories of alternating hard and easy workouts are the basis for the training regimens of virtually every top competitive distance

runner in America today. In the late 1950s, dissatisfied with the track shoes then available, Bowerman experimented in his garage with shoes he made from rubber and kid leather. The Olympic gold medal (which set a world record) for the 400-meter dash in the 1960 Rome Olympics, as well as a bronze medal in the 1964 Tokyo 5,000-meter race, were won by athletes wearing these shoes. (An ironic counterpoint to Bowerman's work is that he now walks with a limp, the result of nerve poisoning from close contact with an experimental glue used to bond soles to uppers.)

One of the track stars recruited by Bowerman in the 1950s was a middle-distance runner, Phil Knight, who went on to attend graduate business school at Stanford. Knight wrote a term paper at Stanford in which he outlined the creation of a successful business selling "track shoes." In 1962, he made good on this proposal by relocating to Oregon and forming Blue Ribbon Sports. (Later the name was changed to Nike, honoring the winged Greek goddess of victory.) BRS became the American distributor for Tiger running shoes, manufactured by the Japanese firm Onitsuka, and in 1964, Bowerman and Knight formed a partnership to sell custom shoes made to their specifications by Onitsuka. Several early Bowerman developments from this phase of their partnership later became standard for running shoes, including the first lightweight nylon upper and the first full-length cushioned midsole.

But Bowerman's most important contribution to running shoes was the waffle sole. He had been experimenting with different sole patterns that would cushion the runner's foot more comfortably and grip the ground more firmly, especially in western Oregon's damp climate. While eating breakfast one morning in 1971, Bowerman was about to take a bite of a waffle when he realized that the shape was perfect for his needs. He took the waffle iron into his garage workshop, poured some urethane into it, and created a waffle pattern that he cut into the shape of a shoe sole. Bowerman first tried attaching his new sole to the heel portions of normally spiked shoes, but at the suggestion of marathon runner Ken Moore he removed the spike plates and created a set of ultralightweight shoes with waffle soles only. These shoes helped the Oregon cross-country team win the 1971 NCAA title.

Bowerman patented and introduced his design to the public in 1972. According to legend (and an early Nike ad), Mrs. Bowerman's waffle iron was preserved for posterity; in fact, it was ruined by the experiment with urethane and promptly thrown into the sanitary landfill across the road from the Bowerman home. "I wonder if anthropologists will realize what they've found when they dig it up ten thousand years from now," he says.

"Boot-In" Ski Boot Carrier
Cantilevered Bicycle Seat

1964–

Inventing is a family affair for the Allsops of Bellingham. It began in 1964 with a single idea: the Boot-In, a carrying and display rack for leather ski boots. The Allsops, father Ivor and sons Jim and Mike, have since invented and marketed devices ranging from shock-absorbing ski poles and a suspension system for bicycles to audio equipment cleaners and office supplies.

In the early 1960s, Ivor ("Buss") Allsop was the manager of the Mount Baker Ski Area, near Bellingham. These were the days of leather ski boots, and Allsop's rental shop had a recurring problem: all the boots came in at once from the slopes, sopping wet from a day's use, and it was impossible to dry them properly by next morning. This resulted in severe curling of the leather at heel and toe. Allsop began thinking about a rack that would hold boots on a shelf and let them dry overnight, hooking the toes and heels of each boot so that tension would keep them flat. He built an experimental rack at the Mount Baker shop, and it worked well enough that he made a single-pair carrier out of plastic, a material then relatively new in the ski industry. Allsop called the carrier the Boot-In, and it was a hit with everyone who saw it. Allsop filed a patent, quit his job, and formed a company to manufacture Boot-Ins from extruded plastic. For years it was the best-selling carrying rack in the world, and over six million have been sold altogether. Allsop later modified the design to make large display racks for retail ski

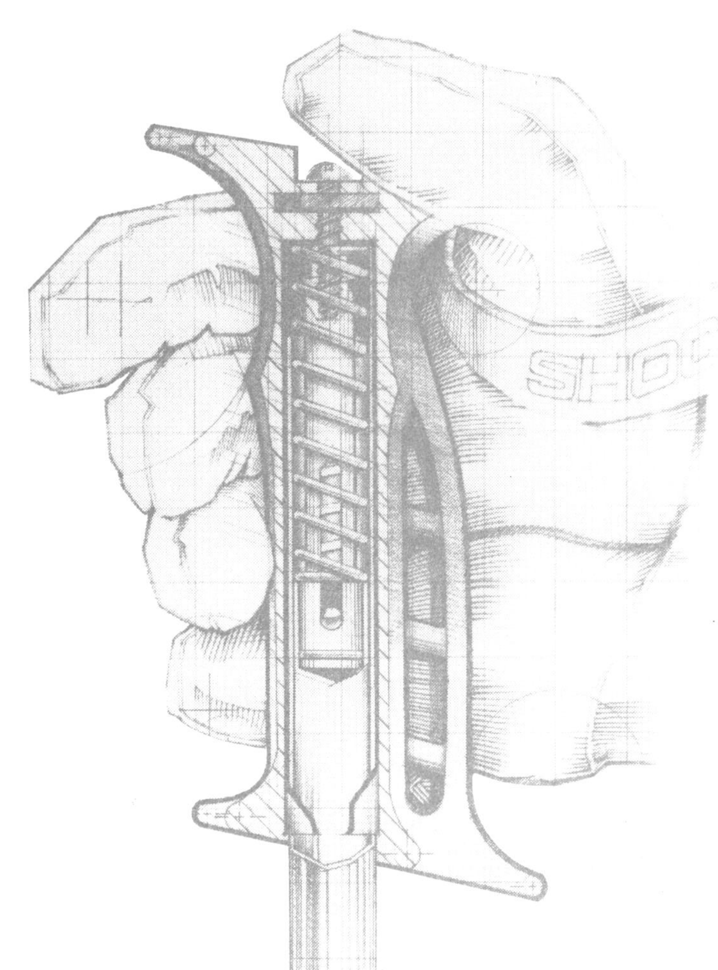

This shock-absorbing ski pole is one of the Allsop family's many outdoors-related products.

shops, and this version also became the world's best-selling unit of its type.

In an era of specialization, the Allsops are noted for their eclecticism. The only common threads in their products are the use of plastic injection molding and their fondness for what Jim Allsop calls "a good, simple idea that no one else is doing." This ability, combined with an instinct for spotting trends, has turned the Allsop Corporation into a multimillion-dollar company with over 450 patents worldwide. Their products range from high-end cassette/compact disc cleaners and computer diskette holders to bicycle racks for automobiles.

In 1989, Jim Allsop patented a shock-absorber system

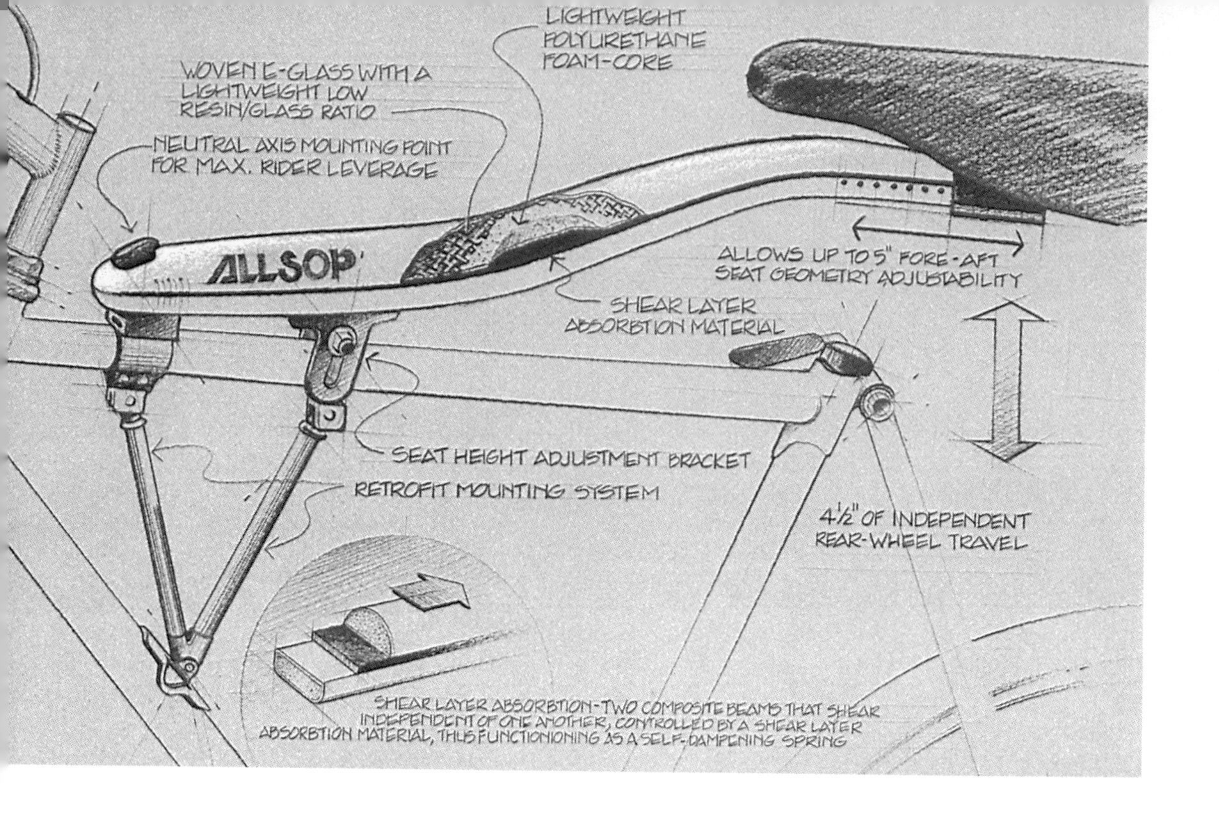

for bicycles, available either as part of a custom bike or retrofitted onto an existing bike. Initial response has been positive among both professional bicyclists and amateur enthusiasts, and the Allsops hope it will prove to be one of the most significant innovations in bicycling since the invention of the derailleur gear in 1905 by Paul de Vivie. Allsop says: "We all had mountain bikes, and it became obvious, just by looking around, that every mode of travel but bikes had a suspension system — even the old stagecoaches." His father adds, "I would say that maybe eighty-five percent of the creative thought on bicycles was done before the early part of this century, and a lot of it was done before 1900. The bicycle was just about the perfect vehicle, and there wasn't much room for improvement until new materials made things like suspension systems possible."

Theirs is not the only suspension system for bicycles, but it differs from other models in that it is not an adaptation of a motorcycle-type setup cushioning the entire bike. Jim explains: "The problem when you're suspending the bike and not the rider is that every time you pedal you activate the suspension system and lose energy. Your pedals have to be lower with a motorcycle-type suspension, too, and pedal height is critical with mountain biking." A system that would isolate and cushion only the rider was the answer. "We tried first to make an articulated seat post that would absorb shock by going up and down. We tried pneumatics and hydraulics, but the various forces at play caused a great deal of friction; it takes a lot of force for a movement to start in each direction, and the amount of energy needed at the top and bottom of each stroke was too much. Also, the rotation of the seat was a problem — any rotation around the axis of the seat post was very objectionable, so we had to solve that."

The Allsops settled on a cantilevered system — that is, with the seat suspended and "floating" free on a beam from the main frame. (Imagine a bicycle seat mounted on a beam extending backwards and up from the front wheel.) Their test beam, mounted on a mountain bike, was made of fiberglass with a shock-absorbing polyurethane foam core. On paper, it made sense to split the beam into two halves, top and bottom. They tried it "whole" first, because it was easier, but during trials it split on its own. Jim says: "Where it split as I was riding it, it was becoming better and better. We made another beam, cut it on the bandsaw and filled the split with urethane. It effectively became two beams, top and bottom, with the halves held together where the seat is attached, and that really improved performance."

The Hacky-Sack

1972

You see people playing with Hacky-Sacks everywhere, in the Northwest and beyond — at outdoor rock concerts, while waiting at ferry terminals, on sunny summer days outside college dorms. The game's popularity on campuses has led one wag to suggest that the official emblem of The Evergreen State College in Olympia be "four guys with ponytails playing with a Hacky-Sack in the rain." It is proof positive that even a small, inexpensive object — say, a leather ball filled with plastic pellets — can be worth a sizable fortune to the right inventor.

The Hacky-Sack is the main ingredient in footbag, a pastime with an ungainly name but with origins in centuries-old games from the Far East. The object of footbag is simple: to keep the bag in the air as long as possible, using only one's feet and legs. Just as there are many versions of Hacky-Sacks, from "official" Hacky-Sacks filled with plastic pellets to homemade versions filled with mung beans or popcorn, there are many versions of the game. Official tournament play has rules similar to volleyball, but footbag can also be played solo or in informal pickup groups, with any number of players standing in a circle.

The Hacky-Sack was conceived by a transplanted Texan, John Stalberger Jr., and a Portlander named Mike Marshall. Stalberger was a college athlete who had suffered a serious knee injury playing football at the University of Texas. Despite several operations and extensive rehabilitation, the knee was still giving him trouble when he visited Portland on a spring break vacation in 1972. In Portland, Stalberger

Hacky-Sack fans like the game's informal spirit, de-emphasis on winning or losing, and offbeat rules (including one that forbids a player from apologizing for a missed kick).

met Marshall, a handyman and fellow athlete, who taught him a "funny game" he played with a beanbag. The game was solo footbag, although it didn't have that name yet. (In solo footbag, a player hops back and forth on alternate feet, bouncing the bag from one to the other.) The game turned out to be perfect for creating a variety of motions to flex Stalberger's injured knee, and Stalberger and Marshall, who called the game "hacking the sack," became regular players. Stalberger settled in Oregon, and the two set to work developing a heavy-duty bag that could stand up to serious punishment.

When Marshall died in 1975, a heart attack victim at age 28, Stalberger decided to carry on the development of the game himself, in large part as a memorial to his friend. He patented the ball he and Marshall had developed (the patent concerns the use of the baseball-style construction for Stalberger's particular type of "foot bag"). When he began demonstrating it to sporting-goods stores and physical education classes in the local schools, the buyers, coaches, and students liked it. Its reputation grew and, in 1983, with over a million Hacky-Sacks already sold, the Wham-O Corporation of California (a keen judge of novelty sports, and already famous for its marketing of the Hula-Hoop and Frisbee) paid Stalberger a reputed $1.5 million for the U.S. and Canadian marketing rights to the toy, which retails for about five dollars. Stalberger's company, Kenncorp Sports of Vancouver, Washington, continues to market Hacky-Sacks to the rest of the world, and Stalberger travels the world promoting both his invention and the sport he helped to popularize.

Racing Stroller

1983

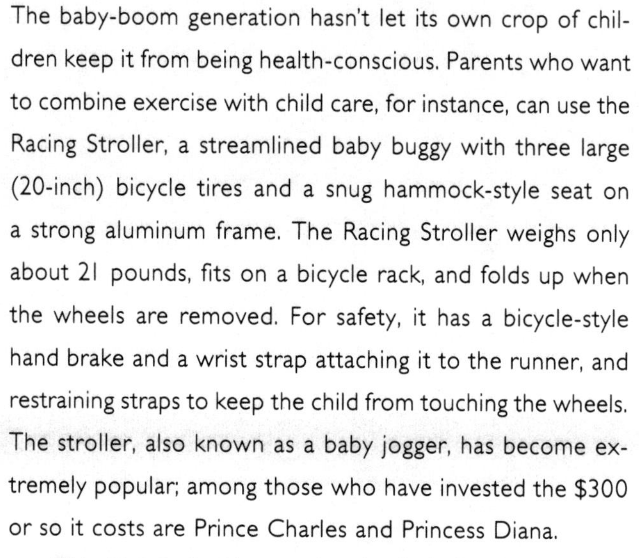

The baby-boom generation hasn't let its own crop of children keep it from being health-conscious. Parents who want to combine exercise with child care, for instance, can use the Racing Stroller, a streamlined baby buggy with three large (20-inch) bicycle tires and a snug hammock-style seat on a strong aluminum frame. The Racing Stroller weighs only about 21 pounds, fits on a bicycle rack, and folds up when the wheels are removed. For safety, it has a bicycle-style hand brake and a wrist strap attaching it to the runner, and restraining straps to keep the child from touching the wheels. The stroller, also known as a baby jogger, has become extremely popular; among those who have invested the $300 or so it costs are Prince Charles and Princess Diana.

Have your kid and keep fit, too: Phil Baechler pushes son Travis in one of his Racing Strollers (above).

This high-tech take on the classic baby pram was invented in 1983 by Yakima's Phil Baechler, a marathon runner who needed a way to train while caring for his infant son, Travis. At the time, Baechler was working the night shift as a copy editor for the *Yakima Herald-Republic*; the only way he could train was by running during the day, which was also the only time he could see his son. His wife Mary suggested that he take Travis along on runs, but bouncing the child in a backpack was dangerous and uncomfortable. Inspired by the sight of covered infant trailers designed to attach behind bicycles, Baechler thought of making a lightweight, sturdy, reliable stroller using bicycle wheels and aluminum. A bicycle racer himself, Baechler already had "a garage full of spare parts and frames" to work with; he bought an old stroller at a secondhand store, stripped it down, and welded on a

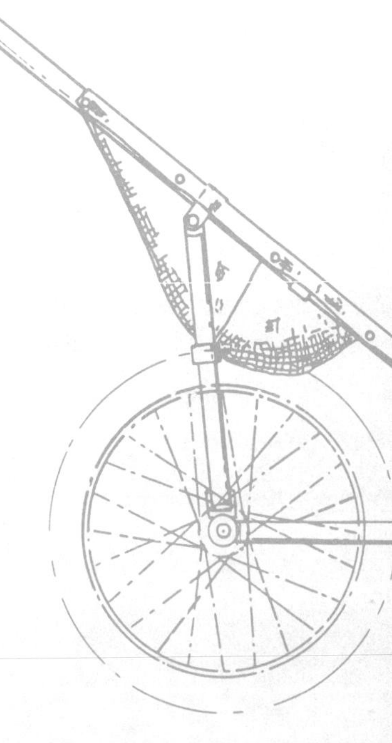

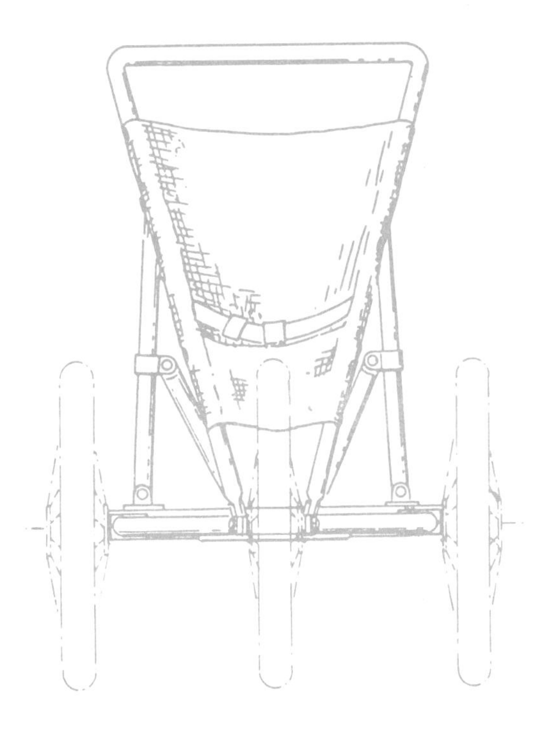

bicycle-style fork and wheels.

"It was really ugly, but it worked. I'd push Travis in this thing, and he'd go to sleep, and people would give me some really curious looks. I took him in the stroller on a ten-kilometer race when he was six months old, and somebody said to me, 'What a good idea!' That was when I started thinking of it for other people. Before that, it had just been a cool toy for me and Travis." (One newspaper profile stated that Travis's first word was "ride"; Baechler says that " 'Mommy' and 'Daddy' may actually have been first, but 'ride' was certainly right in there.") Baechler patented his stroller in 1984, formed Racing Strollers Inc., and, with his wife and his friend Jim Mucklow, began manufacturing them in his garage; three years later, he quit his newspaper job to concentrate full-time on the company, which has since expanded its line to include an all-terrain model, strollers for two or more children, and carriages for disabled children.

Fiberglass Skis

1962

The first successful fiberglass snow ski was tested in 1962 by Bill Kirschner, and the first production model, the Holiday, was introduced to skiers in 1965 by his company, the K2 Corporation of Vashon Island, Washington. Retailing for $80, the Holiday was designed to fill a niche in the rapidly growing recreational ski market roughly midway between inexpensive wooden skis (which retailed for about $50) and expensive metal ones (about $120). K2 skis gained considerable attention when Marilyn Cochran of the U.S. Ski Team won the 1968 World Cup gold medal on a pair of competition K2s, marking the first time the medal had been won on an American-made pair of skis.

In the mid-1930s, Kirschner's brother Don, then 18 years old, badly fractured his leg; it was set by Dr. Roger Anderson, a Seattle orthopedic surgeon (and the father of restaurateur Stuart Anderson). Anderson set Don Kirschner's leg with his own invention, an experimental stainless steel pin-and-splint combination that had been tested only a few times. It worked well, and Anderson later asked the Kirschners' father, Otto, who owned a small manufacturing company, to "take the splint on the road" for him. Unfortunately, the restrictions placed on the use of metal during World War II kept the invention from being marketed and manufactured properly. After the war, looking for new kinds of business, Don thought of creating a modified version of Anderson's splint/pin system for use on dogs and other small animals, since they have considerable difficulty when confined to casts. Bill Kirschner, who is now retired on Vashon Island, recalls: "I had graduated [from the University of Washington, with a degree in mining engineering]. I didn't really want to spend my life in the woods and try to raise a family where it's forty miles to the nearest grocery store, so I decided to join him."

The "liveliness" of a fiberglass ski works like that of a vaulting pole, reacting to the stress of a skier's weight by springing back and propelling the skier forward.

When the idea proved reasonably successful, the Kirschners branched into another, related line: high-quality cages for animals in veterinary hospitals. Bill says: "At the time, there were no cages for animals that were really sanitary. They were all made of metal, with edges and seams that were hard to clean. Reinforced fiberglass was just coming into commercial use then — the boat business is what used it most in the beginning — and we had the idea that fiberglass would make a good animal cage."

In the winter of 1960–61, Bill made himself a pair of fiberglass skis, in part because he "was too cheap to buy myself a pair of Heads," the expensive, high-performance metal skis developed by Henry Head. He tested them that winter at various Northwest locations. The experimental skis performed remarkably well, and Bill was still pleased even after buying a pair of Heads for comparison.

Bill developed and patented a "box construction" method whereby he wrapped unidirectional fiberglass (glass in which the fibers run in the same direction) around a slim core made of spruce that extended along the body, but not the tip, of the ski. The core, he says, "could have been air — you didn't really need it, but you had to have some sort of structure or frame to wrap the glass around." The ski was then put into a pressure mold, and "when you brought it out... it was a usable ski, except for being painted or decorated." This simple, inexpensive "wet wrap" method kept costs to a minimum but produced a strong ski with a great deal of rebound. Bill Kirschner formed the K2 Corporation (named after the two Kirschner brothers, and also for the second highest mountain in the world), and paid $5,000 for the equipment of a California ski firm that had gone bankrupt. Don Kirschner remained as head of their animal-cage business.

Sales were slow but steady after the initial run of 25 pairs; by 1968 annual sales were up to 19,000 pairs. A turning point in the company's growth came that same year when it acquired Chuck Ferries as a sales representative. Ferries, who had been ranked the fifth best slalom skier in the world, was a former Olympic racer and the coach of the U.S. Women's Ski Team. He had also been the Northwest sales representative for Head, and had unsuccessfully tried to interest that company in making a fiberglass ski. During the summer of 1968, Ferries convinced Kirschner to try his hand at making a competition-quality ski.

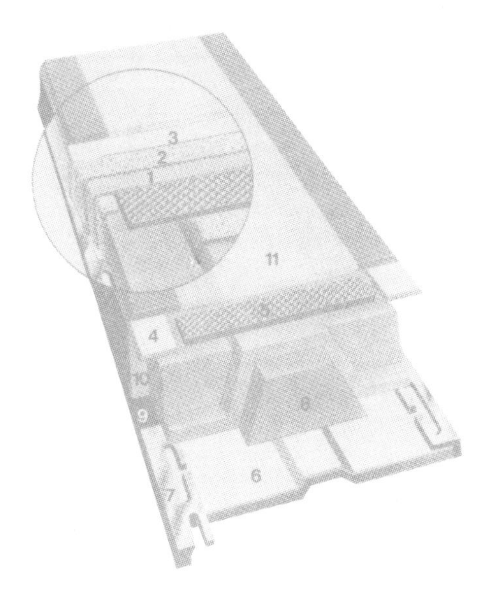

Detail in circle shows appearance of fibreglas box construction before heat molding. Front of cutaway represents appearance after heat molding.

1 Layer of unidirectional fibreglas.
2 Two layers of non-woven fibreglas.
3 Finishing layer of fibreglas surface mat.
4 High-impact abrasion-resistant white ABS top edges.
5 Inertial damping metal foil layer for stability.
6 White superfast P-tex bottom.
7 Keyslot steel edges.
8 Fibreglas-reinforced polyurethane triple core structure for extra strength and light weight.
9 Molded rubber bonding layer.
10 Fibreglas box/rib construction after glas-wrapping and heat molding (see detail).
11 Abrasion-resistant, multicolor solid ABS top.

The basic difference between a recreational ski and a competition ski lies in the amount of torque, the ski's ability to twist. The stiffer the torque, the better the ski will hold on a hard, icy surface and the faster a skier can go. Less-stiff skis have edges that don't "catch" the snow's surface, giving what Kirschner calls "the Sunday afternoon skier" a smoother ride and more control.

Kirschner had to solve the challenge of getting his fiberglass skis to "hold the line" on the icy surface of a race course by controlling the torsional flexibility. Kirschner made several prototypes and asked Robert Albrecht, a structural and architectural engineer at the University of Washington, to use a seismograph under lab conditions to measure the vibrations created by the skis when they hit imaginary bumps. Those skis that most quickly reduced the tremor of their own "earthquake" and yet retained desirable torque characteristics became the basis for K2's racing production models. Kirschner also worked on prestressing skis to critically

dampen their "liveliness." If an entire ski were made with "lively" fiberglass, it would be much too responsive and would fling its skier all over the course; too little liveliness and the ski's performance is dull. Kirschner found the optimum formula was to put "deader" fiberglass at the front of the ski to absorb bumps, and prestressed, livelier fiberglass in the tail. Now, when a racer put pressure on his skis while sitting back and coming out of a sharp turn, the lively part of the ski reacted and accelerated him forward. By the fall of 1968, several U.S. Ski Team members used K2's red-white-and-blue competition skis at preseason training camps, and that same year Marilyn Cochran won the World Cup gold. Since then, K2s have become standard equipment for the U.S. ski teams. Although K2 still manufactures its skis on Vashon Island, Kirschner sold the company in 1969 to Cummins Engine Company Inc. of Columbus, Indiana. It has since been resold several times, and is currently owned by Anthony Industries of Los Angeles. K2 now sells about 400,000 pairs of skis annually, nearly half of them in the United States.

The Conibear Stroke for Crew Racing

1908–16

Pocock Crew Shells

1910s–60s

Hiram Conibear was a professional bicycle racer, team trainer for big-league baseball, and track coach for the University of Chicago before he became an athletic coach at the University of Washington shortly after the turn of the century. Although he had no previous experience in crew racing, Conibear was named head coach for the school's then-nonexistent team in 1904, and in 1908 he began developing a rowing technique that became known as the Conibear Stroke.

Dick Erickson, an oarsman for the UW team who became its coach in the 1970s, says that Conibear's principles, which emphasized leverage created with the legs, have stayed with the sport: "There's no doubt in my mind that the Conibear style of rowing has prevailed. The best crews in the world [today] come closer to the Conibear stroke than any other named style." In Erickson's opinion, newer methods, such as the influential stroke introduced in 1960 by Karl Adam of the West German Rowing Federation, have not stood the test of time: "The West German derivation was based on smaller, stockier people. It's a more powerful, explosive stroke, requiring a higher number of strokes per minute. But the best crews are still tall people, both men and women. In terms of efficiency and leverage, taller people simply do better, and the Conibear stroke was designed for that. Our biggest and strongest muscles are in the legs, and Conibear maximized the pressure or lift created with the drive of the legs. He also shortened the body swing, the amount of body movement in a stroke."

Conibear worked on his technique by borrowing a skeleton from the university's medical school, putting it into a rowing shell, and then moving the skeleton's arms and legs around. He made detailed graphs of the results. The stroke changed over the years as Conibear improved and refined his ideas.

Conibear's stroke was further developed by another legendary figure in UW rowing, George Yeoman Pocock. Pocock, an Englishman, was a third-generation builder of racing shells, and a rower of note who had entered his first professional race at age 17. He and his brother Dick, recruited by Conibear in 1912 to build shells for the UW, incorporated elements of the traditional Thames Waterman Stroke into the Conibear technique. In a memoir written in 1972, Pocock wrote: "These movements are almost impossible to put on paper or explain by word of mouth; they have to be demonstrated by constant practice until the oarsman gets the true 'feel' of the boat...."

STROKE! Hiram Conibear coaching the Huskies. Below: George Pocock with his world-famous rowing shells.

From 1917 to 1922, the Pocock brothers worked for Bill Boeing, building pontoon floats for light airplanes and speed-boat hulls. Then Dick moved to New Haven to build racing shells for Yale, and George returned to the UW to work with Russell ("Rusty") Callow, who had succeeded Conibear as head crew coach. Over the next decades, Pocock and his select group of workmen painstakingly turned out shells that dominated the rowing world and remained the sport's highest standard until the 1970s, when they were superseded in popularity by boats made of synthetic materials. A 1932 *New Yorker* piece called Pocock's work "the finest in the world."

Besides raising the art of traditional shellmaking to an unprecedented degree, Pocock also pioneered several design innovations. Chief among these was the first use of native western red cedar, which Pocock called "the wood eternal," in place of the traditional Spanish red cedar he had used as an apprentice. Western red cedar, used for centuries by North Coast Indians for their canoes, is extremely rot-resistant, and the shells Pocock made with it remained sturdy even after nearly 50 years of hard use. He also developed a shell design with no interior ribs, so as to minimize distortion of the shell's shape, and thus drag from water resistance. Gordon Newell, in his book *Ready All! George Yeoman Pocock and Crew Racing*, writes: "While [Pocock] refused to go into technical details of what was certainly a major trade secret, he explained that the ribless boats were made possible by what he called a 'cedar sandwich,' with fiberglass applied over both the inside and outside of the boat's skin, so thinly as to be absolutely transparent."

Pocock seemed uninterested in making money from shell-building: in the mid-'60s, an eight-oared Pocock craft cost $1,800, only $550 more than the same boat would have cost at the height of the Depression.

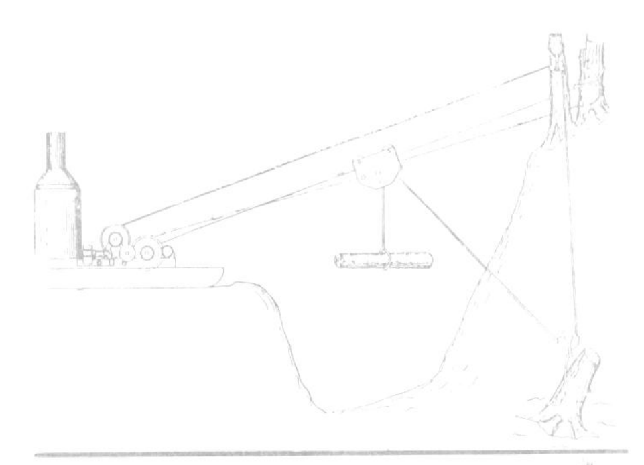

Harvesting Our Natural Bounty

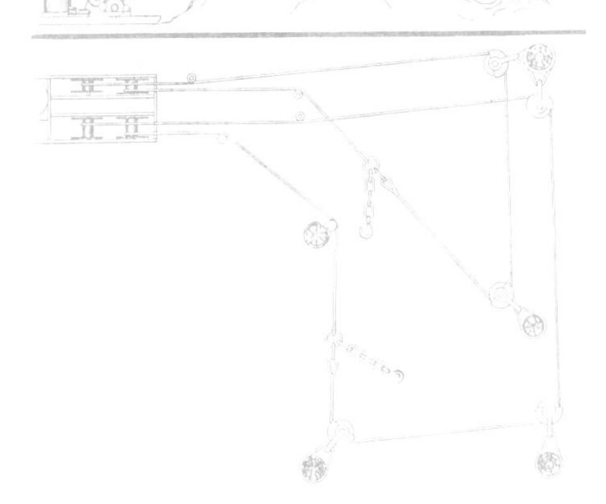

Fish Wheel

1879

The fish wheel, an adaptation of the water wheel, was a device resembling a Ferris wheel, with nets instead of seats, which scooped migrating salmon from the water as they swam upriver. A series of lift nets were attached to a circular frame, which was operated by the current of the river. As fish swimming upstream were scooped up by the nets, they slid toward the axle as the wheel turned and were deposited in a box at the wheel's center. All that an operator needed to do was empty the box as it filled and replace the

An 1897 photo of a typical Columbia River salmon wheel.

occasional waterlogged spar or net.

There were 76 of these perpetual-motion fishing machines in operation along the Columbia alone within 20 years of their introduction in 1879, and each was capable of taking thousands of animals a day from a river — as much as 100 tons a season per wheel. The effect on Northwest salmon runs was so great that fish wheels were allowed to continue operation for only a little more than 50 years. By 1926, they were declared illegal in Oregon, and six years later they were prohibited in Washington. (They are still in limited use, by Native Americans and Native Canadians, on some rivers in Alaska and the Northwest Territories.)

Although the use of fish wheels on the East Coast has been documented as early as 1829, particularly along the shad-fishing areas of the Pee Dee and Roanoke rivers in North Carolina, a patent for one was not issued until 1881. It went to Thornton Williams of The Cascades (now Cascade Locks), on the Oregon side of the Columbia River. A bitter legal battle ensued when another early settler, William McCord, who had seen Williams's experimental wheel the year before, produced a sturdier model and filed a rival patent. The trial, which lasted for several years, awarded the patent to McCord in 1882, when it was demonstrated that Williams's wheel was too similar to an unpatented wheel built in 1879 by still another early settler, Samuel Wilson of Fort Rains, Washington. However, the decision was later overturned on the testimony of 75-year-old Stephen Cole, who testified that brothers William and Robert Thomas, of Brockingham, North Carolina, had been operating their own fish wheel for "50 years or more" on the East Coast. The presiding judge in the McCord-Williams case decided that the fish wheel had long since become public property, and was thus "not a patentable instrument."

The Iron Chink

1903

The Iron Chink — such was its registered name in the casually racist days of turn-of-the-century Seattle — was a mechanized salmon gutter that could clean 110 fish a minute, 55 times as fast as an average human. The first true assembly-line machine for salmon canning, it revolutionized the Northwest's fish-canning and -processing industries and also had a significant impact on the area's social conditions. Whereas Chinese immigrant labor had been used almost exclusively to clean commercial fish catches, now the Iron Chink forced thousands of people to find other forms of work.

It was invented in 1903 by Edmund A. Smith, a Seattleite who had small stakes in the fish-canning and brickmaking fields. Obsessed with finding an alternative to hand-cleaning fish, he worked for months in his waterfront shop before finding the answer. When it came, it was a classic flash of inspiration: he awoke one night at 3 A.M., shouted to his wife that he "had it," and ran to his workshop. When he emerged 10 days later, smiling, Smith promptly borrowed some money and left to see a patent attorney in Washington, D.C.

The Iron Chink, which revolutionized the salmon-canning industry, shown here circa 1906.

In his book *Fisheries of the North Pacific*, Robert J. Browning writes that although the Iron Chink "was not greeted warmly" when the first production model appeared in a Bellingham cannery later in 1903, it quickly gained favor. "Even in an industry as flexible as salmon canning, there was already a body of sacrosanct tradition, although as a fair-sized business salmon canning was barely of age. But tradition or not, the men who ran the industry in those days could read the bookkeeper's work as well as the bookkeeper himself and it did not take most operators more than a couple of seasons to see what the Iron Chink could do for them."

The Iron Chink, which is still used in modified forms in British Columbia and Alaska, works best if the fish are of a uniform size. It can be driven by steam, a common source of power in early salmon canneries, as well as by electricity or gasoline. It can be set to handle fish of different sizes, but it cannot adjust itself from one fish to the next. Browning writes:

> [It] takes a fish fed onto its belt and positions it precisely...[then performs operations including] heading, finning, splitting, gutting and cleaning, while at the same time the fish is being cleansed by water jets. Most of this work takes place in the embrace of two big ring gears called "bull rings." These are three feet in diameter, with 10-inch faces fitted out with pincers, knives and saws.... The fish comes first to the header where a knife flashes down and beheads it neatly. From the header, the fish moves on to the bull ring feed table where tail pincers, one in each ring, clutch it hard while it is being pulled around the rings. As the fish moves, it is grasped firmly by body pincers and held with its back to the rings. A circular saw removes the tail fin; one horizontal knife cuts away the belly fins and a second slices off the adipose and dorsal. The next saw opens the belly and a set of guides spreads the belly flaps. The gutting reel spins away the viscera and the blooder reel with its circular saw slits the membrane over the dorsal artery, cleans out that cavity and, in the final step...a brush proportioned to the shape of the belly cavity wipes away the last of the blood and visceral membrane.

After moving to the next phase, each fish is then given a final wash, molded, cut into cylindrical pieces, and pressed into cans.

The Iron Chink made its inventor wealthy, but his success was short-lived. Smith was invited to display his invention at the 1909 Seattle-Alaska-Yukon-Pacific Exposition, but two days before it opened, on the way to preview the exhibit with his sister, he wrecked their car and it burst into flames. Smith threw his sister to safety, but suffered severe burns himself and died on the day of the fair's opening at the age of 31.

Pak-Shaper

1948–49

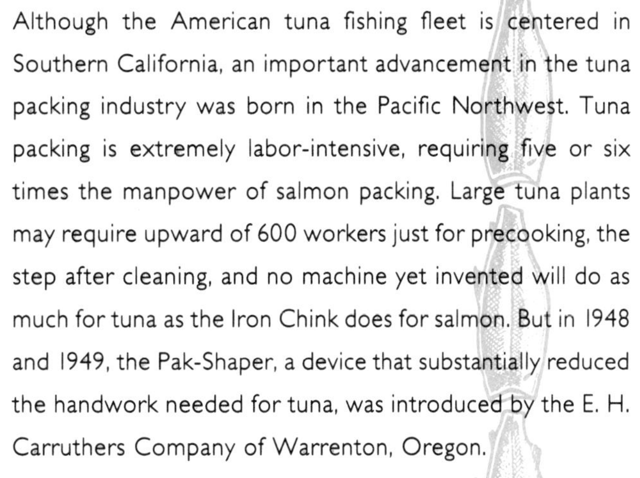

Although the American tuna fishing fleet is centered in Southern California, an important advancement in the tuna packing industry was born in the Pacific Northwest. Tuna packing is extremely labor-intensive, requiring five or six times the manpower of salmon packing. Large tuna plants may require upward of 600 workers just for precooking, the step after cleaning, and no machine yet invented will do as much for tuna as the Iron Chink does for salmon. But in 1948 and 1949, the Pak-Shaper, a device that substantially reduced the handwork needed for tuna, was introduced by the E. H. Carruthers Company of Warrenton, Oregon.

The first Pak-Shaper, installed in the Astoria plant of what was then the Columbia River Packers Association, filled four- or seven-ounce solid-pack cans by molding tuna loins under light pressure into a cylinder of uniform density. It then cut the cylinder into pieces of the correct weight and size, pressed the pieces into cans, and added other required ingredients, such as salt, oil, or spring water. It could perform the work of 50 handpackers with less waste and greater accuracy. The Pak-Shaper filled cans with solid-pack chunk and grated tuna; its cousin, the Pak-Former, was designed to fill only chunk, grated, and pet food cans. Today, nearly all American and many foreign tuna canners use modifications of these early machines.

Fish Ladder

1935

Fish ladders, which allow salmon to migrate upstream stepwise around manmade obstructions like dams, are not a Northwest innovation. But the project that represented the single most extensive use of fish ladders — and a system that incorporated a new twist on the concept, in its use of the "fish elevator" — was the series of waterways built in the mid-1930s to circumvent the large hydroelectric dams on the Columbia River.

Biologists from the U.S. Bureau of Fisheries, including project head Harlan B. Holmes, along with Army engineers and conservation authorities from Washington and Oregon, worked for two years to develop what the Department of Interior called "the first serious attempt to pass large

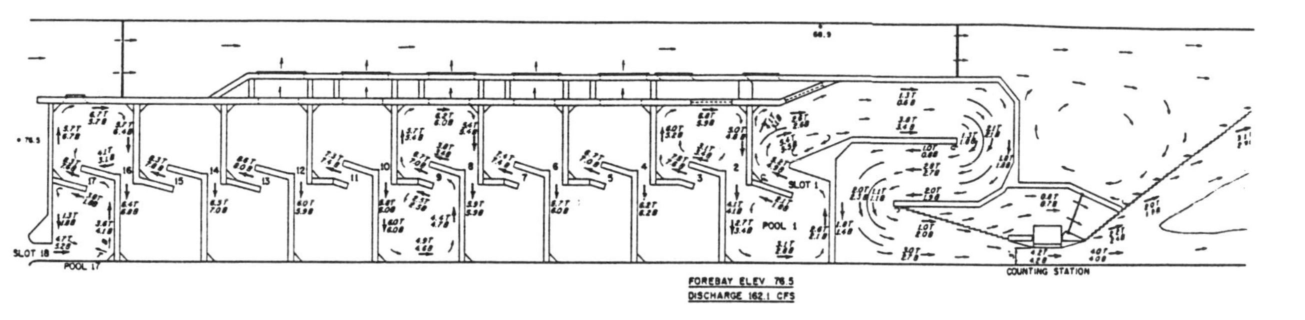

In 1936, Frank T. Bell, U.S. Commissioner of Fisheries, wrote: "A story about stairways for fish, and elevators designed to raise a carload of finny passengers a height of 72 feet, may sound like a chapter out of 'Alice in Wonderland.' On the contrary, these seemingly fantastic devices are being recorded in the history of our own United States."

numbers of migrating fish over obstructions of this size," including the Rock Island Dam, built in 1933 below Wenatchee, the 72-foot dam at Bonneville, completed in 1938, and the Grand Coulee Dam, completed in 1941. The Bonneville portion of the project alone was some 20 times larger than the largest fish ladder then extant.

Since salmon prefer to swim in a swift current, an entrance funnel for the Bonneville ladder was designed to channel the river flow (well ahead of the dam's spillways) into a 1,000 cubic-foot-per-second stream. Migrating fish attracted into the funnel by the fast-flowing water could then navigate a fish ladder, a series of 16-foot pools varying in depth from 5 to 10 feet and connected by flowing water. (Each pool is one foot higher in elevation than the pool below, so that salmon ascended by making the short jump from pool to pool.) Or they could travel through locks. As an article in the March 1936 issue of *Reclamation Era* noted,

> [The fish locks are] the most modern development in fish transportation. These are, in effect, large elevators, 20 by 30 feet. Entrance gates 10 feet square will discharge 200 cubic feet of water per second, as an inducement to the fish to enter the locks. Once the fish are inside, the entrance gates will close, and water willl be admitted through the floor of the lock chamber to raise the water level to that of the top of the dam. The "elevator" itself rises with the mounting water, and when the exit level is reached

the gates are opened and the fish speeded on their way by the device of tipping them out into the river above the dam. Two fish locks will be in operation at the same time. While one lock is ascending the other will be receiving fish below.

At Grand Coulee, neither ladder nor elevator was sufficient to let the fish climb the 500-foot height of the dam. The solution was a collecting trap below the dam; caught fish were transported to holding pools in a nearby canyon, where their eggs were hatched artificially and the fry returned to the river below the dam.

The ladders, elevators, and hatcheries are by no means perfect, and have been the subject of considerable controversy over the years. Recently, a strong effort has been made to place various species of the Columbia River salmon, whose numbers have been devastated by the intrusions of humans, on the endangered species list. As Bruce Brown writes in his book about wild salmon, *Mountain in the Clouds*, "Grand Coulee Dam was…the single most destructive human act toward salmon of all time. When this unladdered colossus was completed in 1942, it closed more than 1,000 miles of spawning rivers and streams in the upper Columbia, killing the famous 'June hog' Chinook that had previously been the mainstay of the great Columbia fishery."

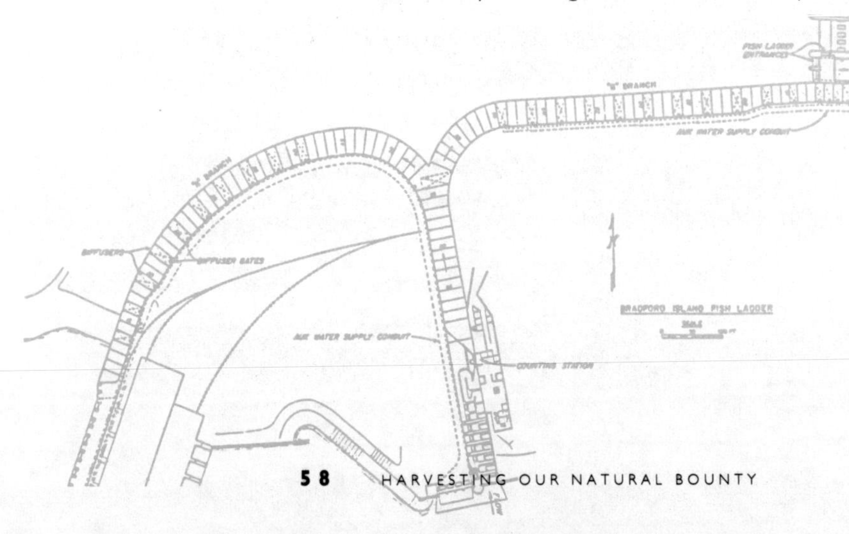

Power Block for Purse Seining

1955

Purse seining is a method of fishing that has been used around the world at least since Biblical times. The method uses huge nets that — at least on today's commercial seiners — can measure as much as a mile long and 600 feet deep, weighing as much as 30 tons. This net, once made of cotton but now usually nylon, has chains or other weights along one long edge (the lead line), floats along the other (the cork line), and a "purse" line strung through rings on the lead line. When a school of fish is located, a small dory is dropped from the boat's stern to pay out the net and the boat turns in a circle to place the net around the fish. When the circle is complete, the purse line is slowly tightened to trap the fish within the net and it is hauled aboard.

The Power Block gave new life to the purse-seining industry. Huge Power Blocks such as this one were custom-made in the late 1960s and early 1970s to accommodate the extra-large floats used on South African pilchard boats.

From ancient times until the mid-1950s, purse seining changed little. Bringing the net in by hand or by the semi-mechanical means called strapping (which involved slowly pulling it in with adjustable cinch-straps) was an extremely difficult and slow process; a complete cycle of paying out and hauling in, called a set, could take up to eight hours if the catch was heavy, and as much as three hours even if the fishermen missed their quarry. The work was so hard, dangerous, and labor-intensive — eight to ten men was a typical crew size on salmon seiners, with more needed for other types — that in many parts of the world the seining industry was in danger of dying out.

In 1955, however, purse seining was revolutionized by the introduction of a simple pulley, the Puretic Power Block, that cut manpower and time by more than half. The Power

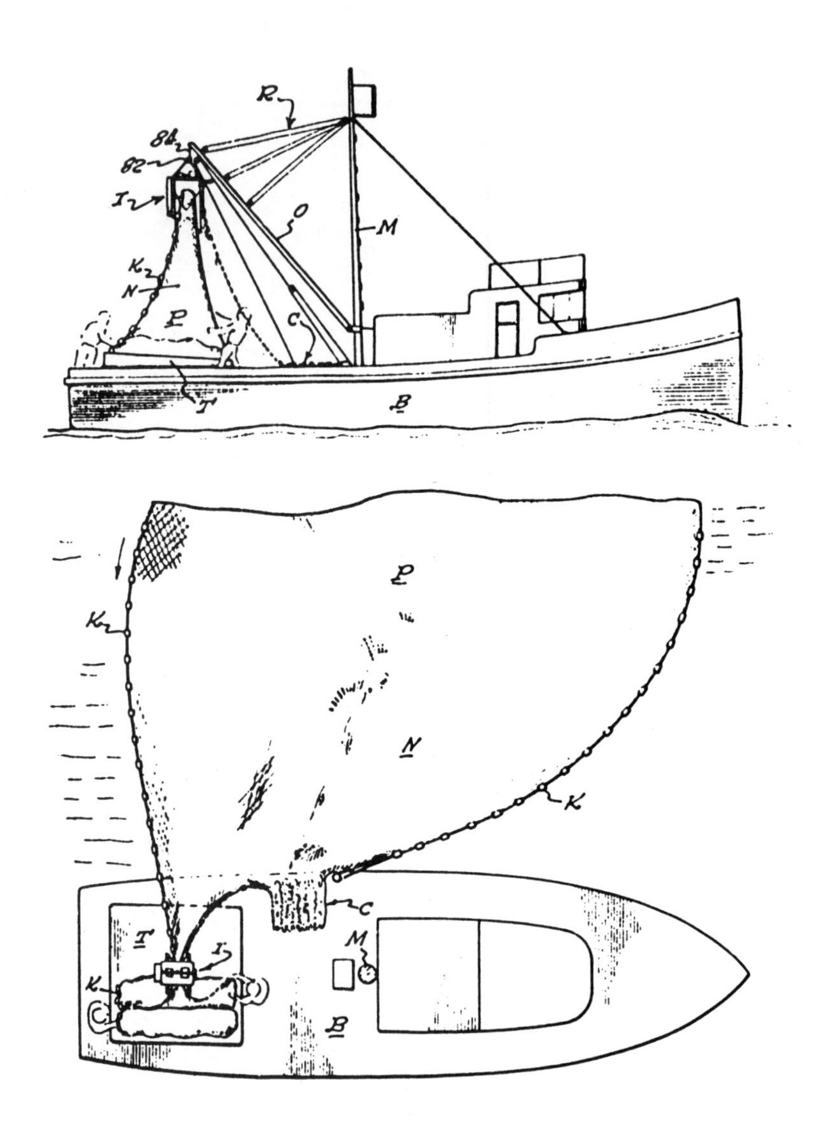

The Power Block is used to haul in nets that can be as much as a mile long and 600 feet deep.

Block's immediate and wide-ranging impact on the fishing industry was a sharp increase in available food supplies — sometimes to the point of dangerous overfishing. It was estimated in the 1960s that half the world's fish, including big-tonnage harvests such as herring, sardines, anchovies, and menhaden, were being caught with Power Blocks; although the percentage is probably less now, they are still used to capture 100% of the world's tuna supply as well as

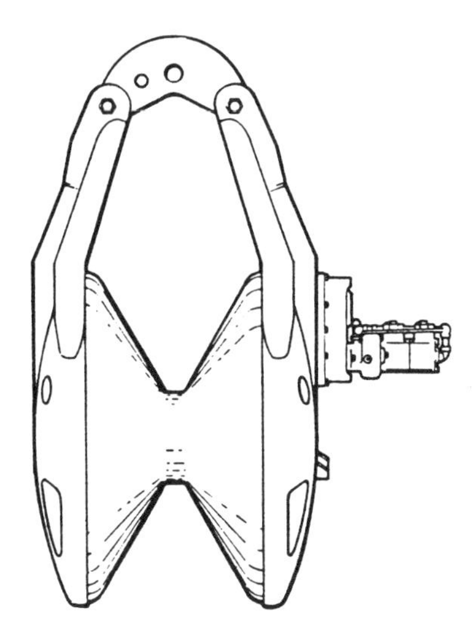

significant portions of the catches of other types of fish. In his book *Fisheries of the North Pacific*, Robert J. Browning writes that the Power Block was "as important to the world's fisheries as was Eli Whitney's cotton gin to the slave owner, as was Cyrus McCormick's reaper to the scythe-weary farmer."

The Power Block is a simple device, a free-swinging aluminum pulley enclosing a V-shaped rubber sheave, or wheel. It is attached to the end of a ship's main boom and powered by a hydraulic motor or by a rope attached to the ship's winch. When the net has been paid out and is ready for recovery, a "messenger" guideline attached to the net is run up into the block and the pulley is activated; as it turns, the end of the net is drawn up through the opening at the top of the V shape, and the net is caught by the rubber sheave. The net can then be pulled on deck with a minimum of slippage and at the maximum speed at which it can be easily stacked by the crew.

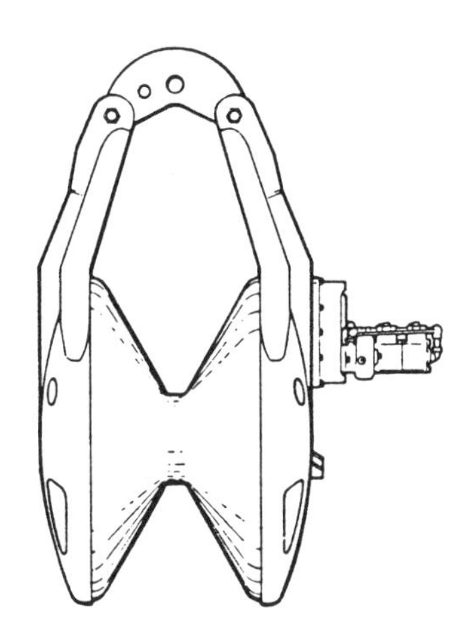

When he invented the Power Block, Mario Puretic (pronounced *pu-RE-tich*) was part of the tightly knit community of Yugoslavian tuna fishermen based in San Pedro, California. He was born in the town of San Martin on the island of Brac; as a young man, his first seagoing experience was on his father's two-masted coastal schooner, which carried wine to ports along the Adriatic. In 1929 he shipped out as a deckhand on a freighter to America and got his start as a commercial fisherman in Seattle in 1937. When the skipper of his boat relocated to San Francisco to fish for sardines, Puretic went along and settled in San Pedro.

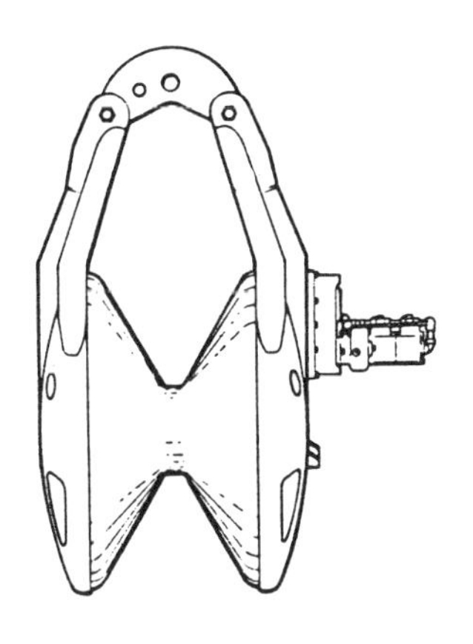

The inspiration for the Power Block came in 1952. While fishing off Costa Rica, the crew was slow in bringing in a set. Sharks attacked the trapped tuna and destroyed both the cotton net and the catch of fish. The boat had to put in for several days while the net was repaired, and

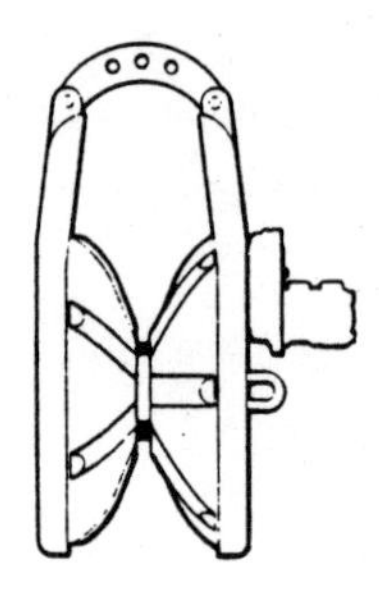

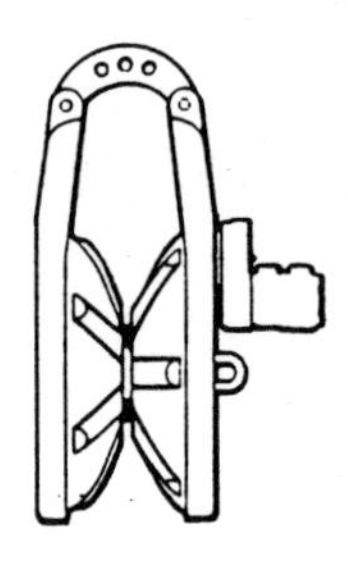

Puretic, thinking about a way to haul nets more quickly, came up with the basic idea for a powered pulley. The fishermen on his boat laughed at the thought, and a San Pedro machine shop that dealt with the fishing community scorned him. Puretic then contacted the Star-Kist canning company, where a lawyer named John Real, who later became the president of Star-Kist, advised him to apply for a patent. Puretic did, and arranged with a metalworking shop in Anaheim to build a prototype. Puretic ran successful preliminary tests, but still could not find an interested manufacturer or fisherman in California.

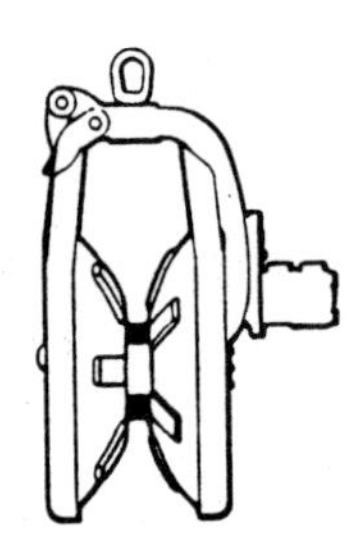

It was the salmon fishermen of the Pacific Northwest who first realized the block's potential. At the same time that Puretic was working on his prototype block, a Seattle company, Marine Contruction and Design (MARCO), was working on the use of large hydraulic drums to haul in seine nets. In March of 1955, MARCO's principals, Peter Schmidt and Don McVittie, were in the midst of installing one of their drum mechanisms in the boat of Paul Glenovich, a Bellingham fisherman, when they received an excited call from Glenovich asking them to stop work immediately. He had just seen Puretic's invention while on a trip to San Pedro and wanted one for his boat.

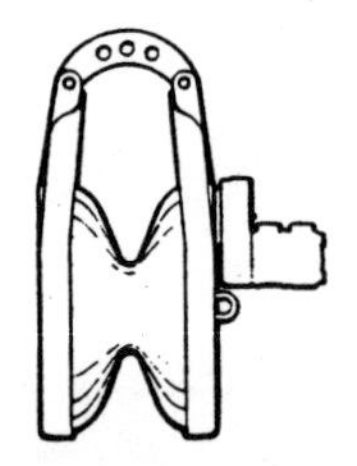

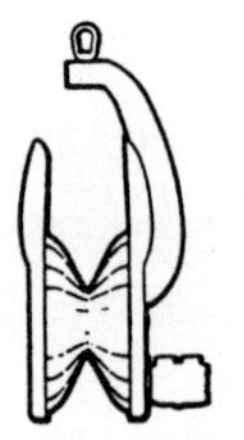

Schmidt and McVittie sensed the block's potential and contacted Puretic. Schmidt recalls, "Mario had this old Dodge coupe with a huge trunk, and he put his prototype block in the trunk and drove night and day to get here. We hung the prototype on the boom of a salmon seine boat, took an old net and made it into an endless loop, and swung the boom over the water. We turned it on and the net went around and around. It worked! *Then* the fishermen started coming, from everyplace. Not just the Yugoslavian fishermen, but the Norwegians and everybody else. We started serving them coffee out in the shipyard, and they'd stand around all day talking about this thing and looking at it." Within two weeks, MARCO and Puretic had a manufacturing agreement — and orders for 100 Power Blocks, to be delivered in 60 days.

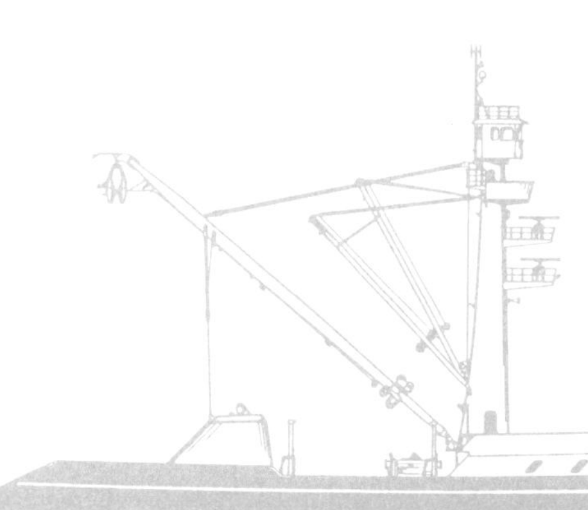

McVittie, Schmidt, and Puretic designed a 150-pound production model made of lightweight aluminum instead of the heavier steel used for the prototype. Although the original idea was to drive the block with a hydraulic pump in the engine room, installing such a complex system in an existing boat would have been costly and difficult. Puretic instead devised an endless rope drive, somewhat like a dentist's drill, that wrapped around the winch head. He recalled later, "The rope drive was a temporary way of motivating the prototype to see whether or not the net would behave as I visualized [it], because even the most experienced of fishermen insisted the net would go crooked. In my original patent, there was no rope drive mentioned, and it was developed...for initial economy. Hydraulic components were very expensive for a new venture and an uncertain business." As it turned out, there was little danger that the net would "go crooked" or be "off the square," that is, that the

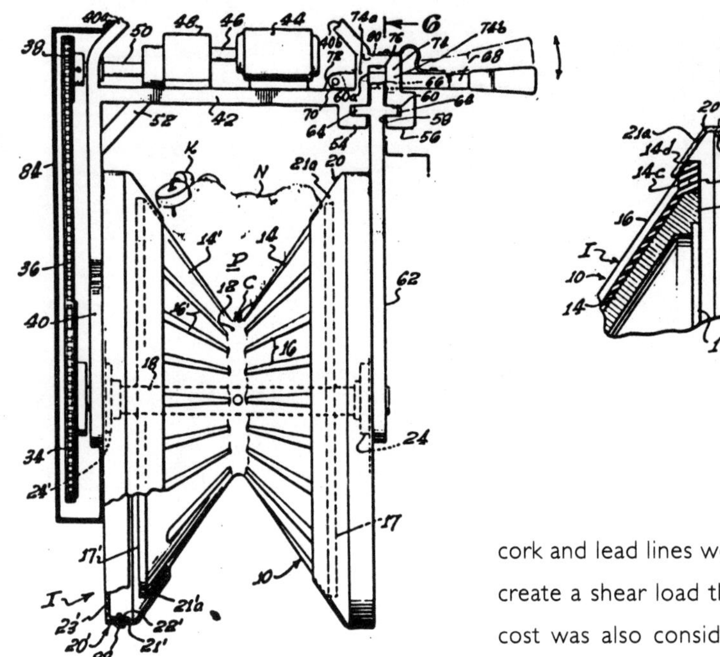

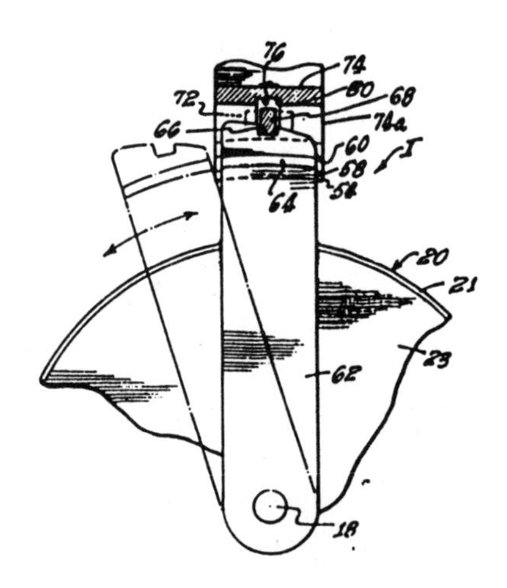

cork and lead lines would be pulled at different speeds and create a shear load that could rip the net. The block's unit cost was also considerably less than first feared; the first Power Blocks sold for $1,250 apiece, with virtually no installation fees, in contrast to the drum seining equipment MARCO had been experimenting with, which required expensive modification of a boat's deck.

The first production Power Block was bought by Seattle fisherman Gordon Nelson for his seiner *New Sunrise*, and by the middle of June 1955, MARCO had delivered all 100 of its orders. The block was an instant success. Soon after its introduction, the Power Block was adopted by the California tuna fleet that had once rejected it, and from there went on to worldwide use. MARCO is still the world leader in Power Block manufacture; in 1975, Mario Puretic, living in retirement in Florida, was presented by the U.S. Patent Office with its Inventor of the Year award.

Spar Logging

1914

Oscar Wirkkala uses his bedroom as a workshop for his propeller designs. Below: Wirkkala is on the far right in this 1914 picture of his logging engine.

Spar logging, a landmark improvement in the speed and efficiency of timber industry operations, was invented by Oscar Wirkkala, a Finn who had emigrated to Washington State in 1898. The method involved hanging an overhead line with blocks from a "spar tree," selected for its height and favorable location with regard to the trees to be felled. A logger known as a high-climber ascended the tree, cutting off branches until he reached a point 175 to 200 feet up and about 50 feet from the treetop; then he cut off the top of the tree. Next came the rigger, who attached pulleys and cables to this makeshift crane and installed a loading boom about 20 feet off the ground. The system was put in motion when a cable with grab hooks on the end was carried to the log to be skidded; by means of a steam or gasoline skidder, the log was dragged within reach of the loading boom. The boom then was used to raise the log off the ground and load it on a nearby vehicle or to stack it for future loading.

Wirkkala patented the process and hardware for spar logging in 1914 and subsequently became a kind of Johnny Appleseed for the new method. His daughter, Margaret Kallio, recalls: "He had a God-given ability to see how things worked, how they should go, and he felt that he should use it to benefit others. So he traveled from camp to camp,

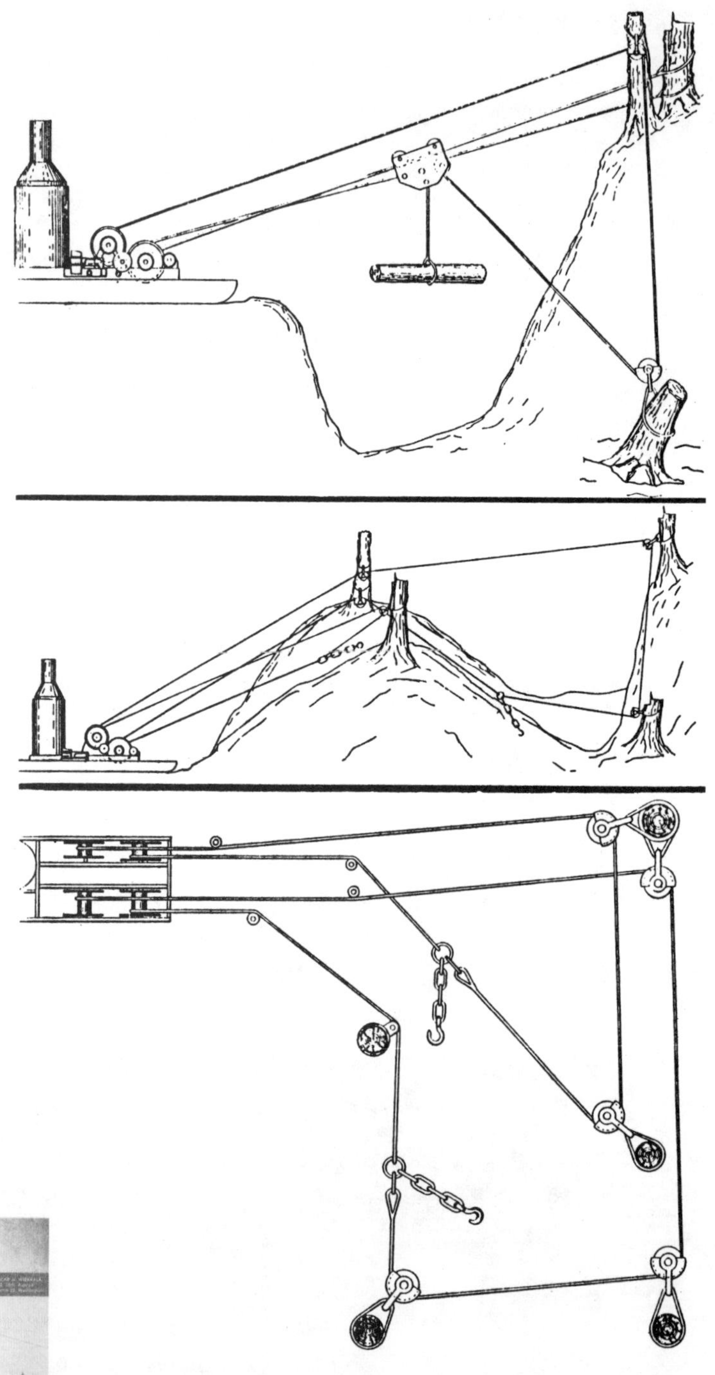

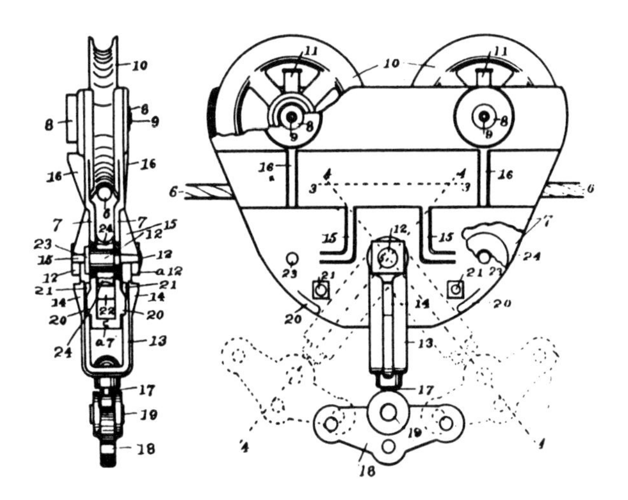

Wirkkala's "skyline equipment" for spar logging was extremely effective at moving the giant Douglas firs of the Pacific Northwest.

just for the regular wages they were paying, showing people how to do these things."

Before his death in 1959, Wirkkala was responsible for several other logging-related inventions, including bull and choker hooks, flexible-strand wire cable, a donkey engine, and a lubricating system for blocks. (Another daughter, Elizabeth Gentala, remembers that he used to cut his designs for new logging hooks out of rutabagas in the kitchen of the family's apartment in Seattle.) Wirkkala also worked on designs for several other projects, including a plywood factory, gold-dredging equipment, a multiengine airplane, and an improved boat propeller. He refused to be slowed down by the fact that he could barely read or write English. His wife translated correspondence, his two daughters served as secretaries, and his younger brother Sam handled the accounts. Wirkkala did, however, keep personal notebooks, a charmingly mixed-up amalgam of Finnish and English in which Seattle is "Seeatli," Ilwaco is "Ilvaako," and the Lewis & Clark Shipyard is "Luuse Kurk Siipjaart."

Plywood

1904

Plywood came into its own as a major element of wood-based construction when the early loggers of the Pacific Northwest found that the soft wood of the Douglas fir was perfect for it. Douglas fir ply was used first for door panels, and the industry began to grow rapidly after a few hand-made panels were shown at the Lewis & Clark Centennial Exposition in Portland in 1905. The primary source of Douglas fir plywood for many years afterward was the Portland Manufacturing Company of St. Johns, Oregon. The technology was still relatively primitive: glue was applied by hand, wooden presses with hand screws were used to hold the ply panels flush until the glue set, and the wood was dried in ordinary lumber kilns. The first plant solely built for and devoted to Douglas fir plywood was probably the Elliott Bay Mill Company in Seattle in 1920; by then, much of the production work had become mechanized.

Plywood was used mostly for the interiors of houses, because the casein glue that formed it (made from the whey of milk left after the protein is removed) was not moisture-resistant. In the early 1920s, Irving F. Laucks, a Seattle assayist, became intrigued with a soybean cake from Asia that was being imported to America as animal food. Laucks developed and patented a moisture-resistant soybean glue (disparagingly called "bean soup" by some members of the industry) that became a standard product until it was replaced by truly waterproof resin glues in the late 1930s.

Chain Saw

1920

Chain Saw Chain

1935

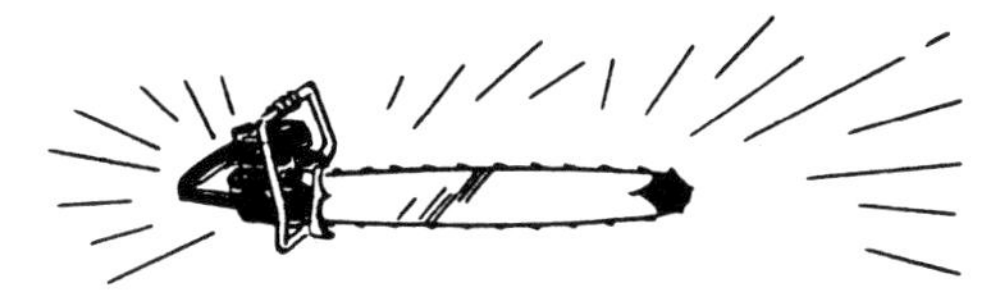

The tool that has had perhaps the single most important impact on logging — the chain saw — was a long time coming. At least a dozen patents for various impractical versions were filed before the turn of the century, but the first practical portable chain saw was not patented until 1920. The Wolf Electric Drive Link Saw, made by Charlie Wolf of the Peninsula Iron Works in Portland, was powered by a small gasoline generator. It featured a chain that could operate in either direction, so that the chain could be reversed as it became dull; it bucked less and was generally easier to handle than any previous chain saw, and came in three sizes ranging from 70 to 90 pounds.

The logging industry was slow to accept it, fearing the loss of jobs it would create, and it did not become a commercial success.

Another Portlander — Joe Cox, an Oklahoma-born mechanic and general handyman who had migrated to Oregon in 1931 to work as a logger — became interested in improving the performance of chain saws. Around 1935 he began to focus specifically on the mechanics of the chain itself. Previously, power saws had operated on the crosscut principle, a rough method with a great deal of waste; and crosscut

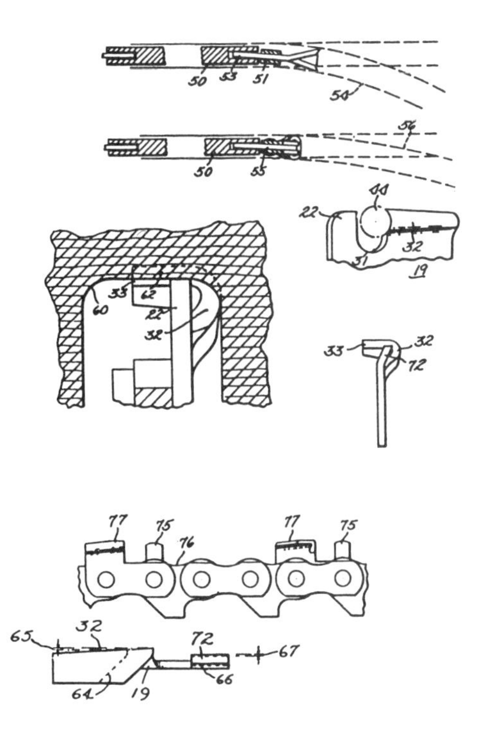

Joe Cox's saw chain cut through wood cleanly and smoothly, using the same back-and-forth principle that comes naturally to the timber beetle.

chains became dull quickly. While chopping wood for his own house, Cox noticed the larvae of the timber beetle, *Ergates spiculatus*, and how the beetle cut through wood: instead of scratching or burrowing directly ahead, the grubs used the two cutters on their heads to make left-right crosswise motions. They worked extremely fast and produced almost no waste. In the basement of his house, Cox designed and built a chain based on the timber beetle's actions, incorporating two cutters on a single head. He patented his invention, but it did not enter the market until 1947, when Cox formed the Oregon Saw Chain Company (later Omark, for "Oregon trademark," and now a part of the Blount Company). Cox's chain was an immediate success, and most saw chains in use today incorporate a modified version of his design.

Logging Tools and Techniques

1881–1970

The term "skid road" (sometimes misstated as "skid row") today refers to a derelict, seedy part of town. Originally, the term referred to an important innovation in the history of Northwest logging. Skid roads were created by clearing tall trees and felling smaller ones crosswise every few feet down a slope, making a path down which the larger trees could be skidded. (Another timber-related term coined in the Northwest is the word "litterbug," originated in the 1940s by Arthur W. Priaulx as part of a campaign to raise public awareness about the need to keep forests green.)

A number of low-tech but effective tools and techniques were developed specifically to log the giant Douglas firs of the Pacific Northwest, where the climate, terrain, and size of trees presented a new challenge to Eastern loggers. Among the tools developed by anonymous Northwest lumbermen before the turn of the century were the massive felling saw, the equally powerful felling ax, the springboard (used by loggers to climb above the swelled portion of an otherwise straight Douglas fir), and the splash dam, used to turn small streams into rivers large enough to float logs. Dams made of logs, cross-tied and bound with wire, were used to form ponds, which would then be filled with logs; when released, the dammed-up water would sluice the logs downstream. The first use of the splash pond in the Northwest may have been by logger Alex Polson in Pacific County, Washington, in 1881; within 10 years, use of the splash dam was standard practice in Douglas fir country.

Balloon logging was first proposed in 1917, by logger

R. H. Barr of Castle Rock, Washington, who suggested that surplus zeppelins be used to haul timber from otherwise inaccessible locations. In the early 1970s, Wes Lematta of Columbia Helicopters in Portland pioneered the use of helilogging, and in the 1980s the Aerolift Company of Tillamook, Oregon, synthesized the two by developing a 178-foot "cyclocrane," a cross between a helicopter and a blimp, to reach remote timbering locations.

In 1935, Gifford Pinchot, chief of the Forest Service, made a famous statement: "Wood is a crop, forestry is tree farming." In the summer of 1941, in Montesano, Washington, Governor Arthur B. Langlie dedicated the first bona fide tree farm: 120,000 acres known as the Clemons Tree Farm. The concept of farming trees had a sweeping impact on the forestry world; by 1981 Washington State alone had 615 tree farms, covering nearly six million acres of land, and tree farming had spread to 81 million acres nationally. One of the tools developed to facilitate this massive effort to reseed forestland was a "double-barreled shotgun" used to plant Douglas fir seedlings. Invented in the early 1970s by Philip Hahn, a forester for Georgia-Pacific, it halved the time previously needed for crews to plant seedlings. One barrel of the "shotgun" incorporates a dibble, a tool used to punch holes in the ground for seeds; the other barrel has a mechanism that lifts the seedling from its storage container, complete with the bullet-shaped compacted soil surrounding its roots, and drops it safely in the hole.

One by-product of timbering is the Pres-to-Log (above), invented in 1929 by Weyerhaeuser engineer Robert T. Bowling as a means of recycling wood waste.

One of the most familiar of the dozens of lumber by-products originating in the Northwest is the Pres-to-Log, invented in 1929 by Robert T. Bowling, chief engineer of the Clearwater Timber Company, a division of Weyerhaeuser. These wood-fuel briquettes are manufactured from sawdust, splinters, and chips formerly burned as waste. Bowling found that a combination of pressure, moisture, heat, and cooling would compress the wood waste (which has a cellular structure that makes it resistant to direct pressure) into a very compact form that burned slowly, with a high degree of heat and a negligible amount of smoke.

Household Wonders and More

Self-Tipping Hat

1896

The self-tipping hat proves conclusively that the Victorian-era obsession with manners could reach even relatively remote outposts like Spokane. In 1896, two residents of that city, James C. Boyle and John Neill, patented their "Saluting Device," a complex, windable system of gears and counterweights, set inside a hat and clamped to the wearer's skull. As Boyle stated in his patent application, this formed

> a novel device for automatically effecting polite salutations by the elevation and rotation of the hat on the head of the saluting party when said person bows to the person or persons saluted, the actuation of the hat being produced by mechanism therein and without the use of the hands in any manner.... To carry into effect the broad feature of this invention, which comprehends the automatic elevation and rotation of a man's head to effect a unique salutation, I preferably employ said mechanism held in a case removably clamped on the head of the wearer of the hat, while the hat is detachably secured to the working parts of the device that raise the hat, completely rotate it and deposit it correctly on the head of the wearer every time said person bows his head and then assumes an erect posture, all parts of the novel device being completely inclosed [*sic*] and concealed by the hat.

To properly salute a lady, in short — even if one happened to be riding a bicycle or with the hands otherwise

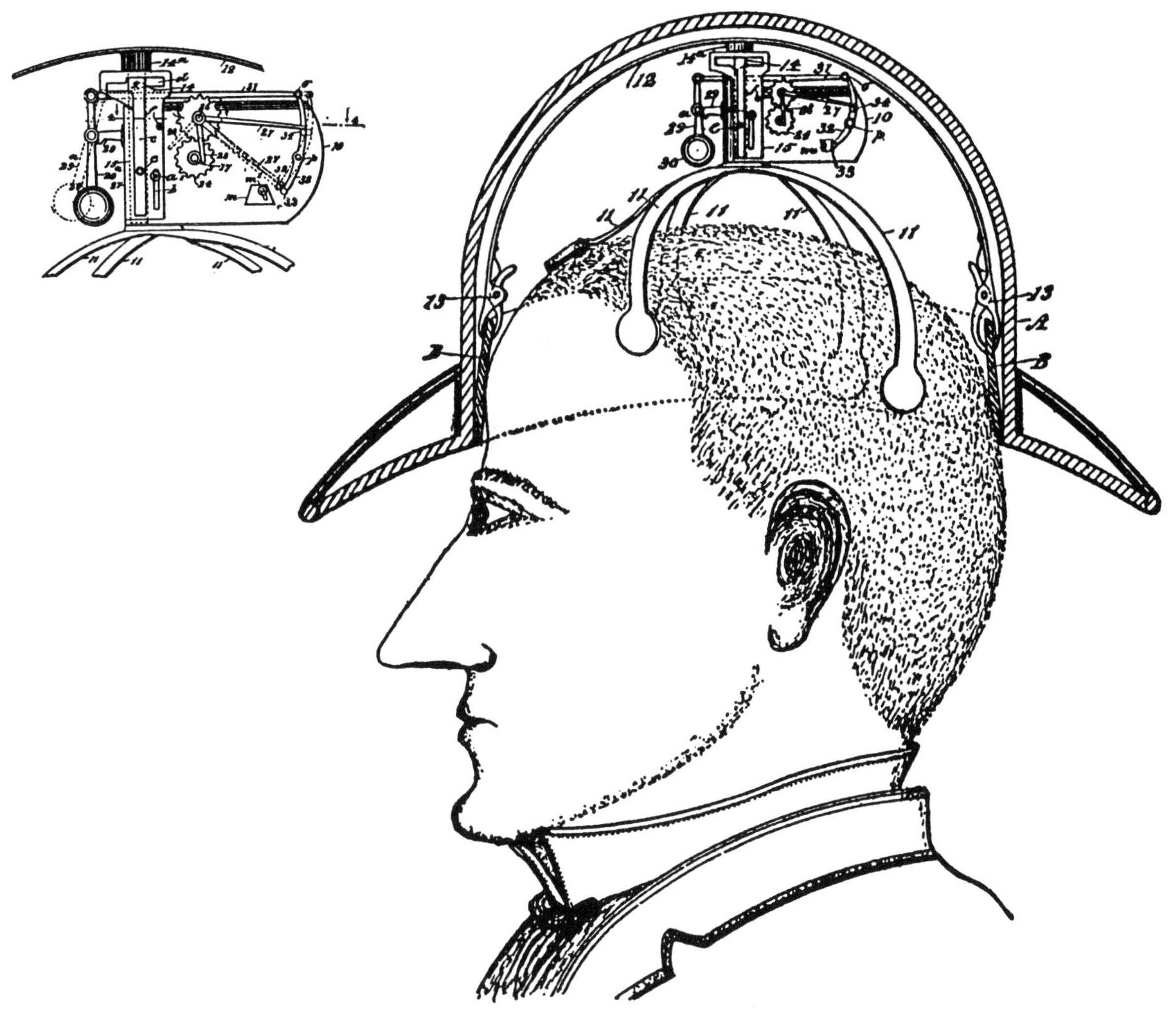

occupied — all that was needed was to lean forward. The hat would tip itself, rotate completely around, and snap back smartly onto the head as the wearer stood up straight again. Boyle was apparently mindful of the hat's possibilities as a novel advertising gimmick, since he stated so in his patent application, which brings up the possibility that his tongue may have been planted at least partly in his cheek when he and his partner created their proposal. Alas, no model was submitted with the patent application, the law requiring this having been abolished in 1880 because the Patent Office was running out of space.

The Oregon Boot

1866

Warden J. C. Gardner of the Oregon State Penitentiary patented the chilling and highly effective penal device known as the Oregon Boot. (This was the same year, 1866, that the State Pen was moved from Portland to Salem.) The Boot was an extremely efficient escape deterrent, especially in prisons that, like many in the West at that time, lacked high walls. Essentially a more compact version of a ball and chain, it consisted of a heavy iron ring clasped around a prisoner's ankle with an iron strap that ran under the foot and a padlock to secure it. The Boot weighed anywhere from 5 to 28 pounds; each prisoner wore his day and night, a sure guarantee that no one would stray far. It was copied widely and used extensively for many years in the Western states and territories, first as a round-the-clock measure for all prisoners, then as a threatened punishment, and finally only for transportation purposes. Even as late as 1939, history records that a prisoner at Mill City was "ironed out" while in transit.

Not all corrections officers approved of the device. In 1873, Superintendent W. Watkins noted that "a great wrong we are compelled to put on our prisoners for want of sufficient walls is the Gardner Shackle.... There are prisoners who have worn this instrument of torture, known inside the prison as a man-killer, until they are broken down in health and constitution.... Men lie in hospital for weeks from wearing these things, suffering great pain and begging to be relieved from the load.... It is murder, and of the worst type." Its use began to taper off soon after, especially when the warden who succeeded Gardner wrote to then-Governor Stephen

Chadwick: "Such punishment can not lead to reformation...."

Examples of the Oregon Boot can be seen at the Oregon Historical Society Museum in Portland, and at the Old Penitentiary Museum in Boise, Idaho.

ANTI-SNORING DAM

In 1953, Elsa L. Leppich of Seattle patented a plastic dam, to be worn while sleeping, that fit between the lips and teeth; by blocking the passage of any breath or moisture, it forced the sleeper to breathe through the nose. The patent claims that the devices "have been found to be an aid to hearing by correcting some causes of deafness."

AERO Alarm

1907

Firecracker Alarm

1966

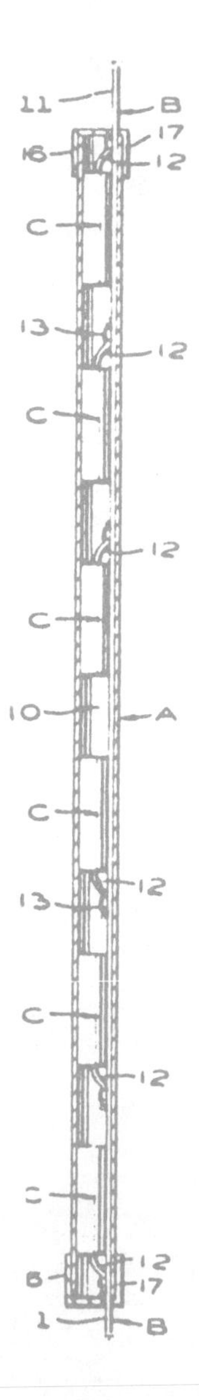

George L. Smith, a Scotsman living in Seattle in 1907, patented a fire alarm system which he marketed, primarily to office buildings, under the name AERO Alarm. It used pneumatic air tubes that, when pushed out of shape by heat, activated a bell, in much the same way that the rubber hoses we drive over in modern gas stations call attention to our presence. Smith's system called for the installation of air-filled copper tubes along the ceiling or high on the walls of the protected area. Heat from a fire would cause the air inside the tube to expand, pushing a diaphragm at the end of the tube. This tripped a switch and activated the alarm.

The main problem with the AERO system was that normal barometric and temperature changes in the atmosphere often caused false alarms. In conjunction with ADT, the New Jersey–based company that bought out AERO in the 1920s, Smith worked during the 1920s and '30s to overcome this, eventually settling on a highly sensitive adjustable vent which compensated for normal pressure and temperature changes. The alarm was also improved by changing it to operate on a rate-of-rise principle. Early alarms used a fixed-temperature system; they were automatically activated when the temperature exceeded a certain, pre-set level. More sensitive rate-of-rise alarms are activated when room temperature increases at an abnormally fast rate, which is generally at a much earlier stage of a fire.

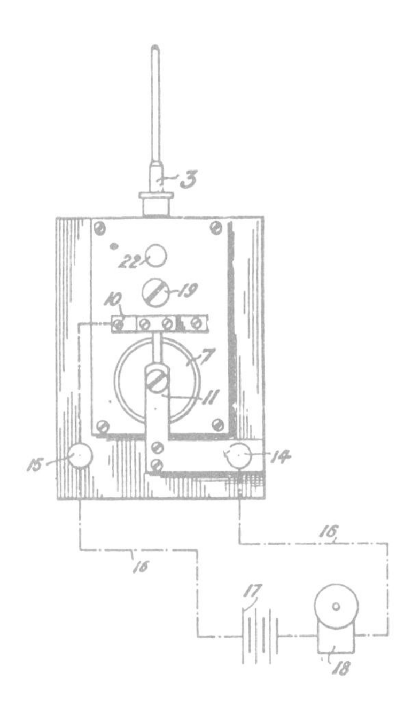

The AERO alarm was successful financially until it was superseded by electronic alarms.

A design from the Northwest that did not fare as well was the firecracker-powered alarm patented in 1966 by Raymond R. Richards of Fall City, Washington. This device consisted of a tube with an exposed fuse, a string of firecrackers spaced at intervals inside (or, as the patent application put it, "a plurality of firecrackers disposed in the bore"), and caps at the ends to prevent children from tampering with the alarm. The theory was that heat would cause the firecrackers to explode, alerting occupants of the imminent danger and giving them enough time to escape. It is not clear why the inventor felt that exploding firecrackers would not add significantly to the danger level of a burning house.

Dick, Jane, and Sally

1940

It is estimated that over 20 million American children learned to read with the help of Curriculum Foundation Readers, better known as the Dick and Jane books. These textbooks, introduced in 1940, became one of the most successful reading series in the history of publishing and helped Scott, Foresman & Company of Illinois to become the largest publisher of school textbooks in America. The series' general editor was William S. Gray, a distinguished professor of education from the University of Chicago, but the person who wrote the books — the "mother" of Dick, Jane, and Sally — was a former first-grade schoolteacher and West Seattle

SALLY DOES IT
BY
DOROTHY WALTER BARUCH
AND
ELIZABETH RIDER MONTGOMERY
ILLUSTRATIONS BY
ROBB BEEBE
D. APPLETON-CENTURY COMPANY
INCORPORATED
NEW YORK LONDON
1943

Baby Sally, Dick and Jane's adventurous younger sister, was based on Janet, the daughter of schoolteacher-turned-writer Elizabeth Rider Montgomery.

resident named Elizabeth Rider Montgomery.

Dick and Jane, their baby sister Sally, Spot the dog, Puff the cat, and Mother and Father were as familiar and influential as beloved childhood toys to young readers in the 1940s, '50s and, to an extent, the '60s and '70s. Although the books later came under attack as being sexist and racially and culturally one-sided (a 1972 article by a National Organization of Women task force was entitled "Dick and Jane as Victims"), their bland, suburban settings did much to shape the social values of the youth of Sputnik-era America.

The Dick and Jane books represented a watershed in the evolving ways we teach our children to read. They used the basal system, which advocates the introduction of entire "sight" words, one at a time, to teach a base of vocabulary; the simple, repetitive narratives ("See Dick run! Run, Dick, run!") were composed of words already in the reader's spoken vocabulary. Questions at the end of each chapter tested comprehension. This system has largely fallen out of favor. Phonics, the sounding-out (or "deciphering") of letters and groups of letters, is the dominant system today, although the basal reading technique still has advocates.

Dick and Jane also marked an important transitional period in the content of children's books, since they were an important element in the shift toward a comprehensive systems approach to reading that incorporated elements of the students' real lives. In the 19th and early 20th centuries, texts for teaching reading made heavy use of Biblical references, fairy tales, and other topics that were distantly, if at all, related to the everyday lives of children. In the 1920s and '30s, this norm was superseded by a system that used specific reading texts reflecting everyday life and nontextbook volumes to teach specific subjects such as math or geography.

For Elizabeth Rider Montgomery, the question was less

pedantic. She simply wanted to write books that wouldn't bore children. (Some who learned from them might say that Dick and Jane were boring, but they were certainly an improvement over their predecessors.) Born in Peru to missionary

parents, Montgomery grew up in Missouri, graduated in 1925 from Washington Normal School in Bellingham (now Western Washington University), and taught first grade in Seattle, Aberdeen, and Los Angeles before marrying in 1930 and retiring to raise a family there. Montgomery had frequently been frustrated with the lack of high-quality materials available to reading teachers. In a *Seattle Times* profile in 1972, she said: "You should have seen the material we were using then. The readers were strictly phonics, no stories, the cat-sat-on-the-mat type of thing." She began writing first-grade primers the year after her marriage and had produced 100 books without finding a publisher when, in 1938, one of her manuscripts came to the attention of a Scott, Foresman editor and she was hired as a staff writer at a salary of $100 a month.

Scott, Foresman editor Zerna Sharp suggested the basic characters of Dick and Jane, but Montgomery developed them fully and created the character of baby Sally entirely on her own, inspired by her own infant daughter, Janet. "I got my story ideas out of my own life. My daughter was just at the age where she was always getting into things and asking questions, and I based many of the experiences in those first books on her antics."

Sally Does It was Montgomery's first published book, in 1940, and she followed it with three books published that same year: *We Look and See*, *We Come and Go*, and *We Work and Play*. By 1946, when Montgomery returned to the Northwest from Los Angeles and settled in West Seattle, her characters were familiar to young readers across the country. She continued to write for Scott, Foresman until 1963, when she began writing nontextbooks, juvenile books including biographies of Chief Seattle, Henry Ford, and Dag Hammarskjöld.

Kirsten Pipe

1936

Wind tunnels, air-raid sirens, fire extinguishers, smoking pipes, standard-voltage neon lights, anti-bug lamps, and innovative propellers are just a few of the nearly 100 patents issued to the prolific Frederick Kirsten.

Kirsten left his native Germany in 1902 as a 17-year-old cabin boy on a three-masted trading schooner, intending to become a sailor. Eighteen months later, after a horrendous trip across the Atlantic and through the Panama Canal that included becalmings, near-mutinies, scurvy, and the death of a captain, he jumped ship in Tacoma. He spent his first night in America hiding in the woods, convinced, thanks to his knowledge of James Fenimore Cooper, that wild Indians lurked behind every tree. Kirsten found work as a farmhand, and began displaying his knack for invention by devising a stump-puller for his boss. He earned enough money to put himself through the University of Washington's school of electrical engineering, was appointed to its newly established faculty of aeronautics in 1915, and throughout his long career was instrumental in the department's development as a major aeronautical engineering center. But it was his gift for creating nonflying gadgets that made him both well-known outside academia and well-off financially.

The most successful of his inventions was the Kirsten Pipe, still in production and still considered "the Cadillac of smoking pipes." In 1936, Kirsten's doctor ordered that he give up cigarettes. A self-described "inveterate smoker" since age 11, he turned to wooden pipes, found them lacking, and designed a pipe with several improvements, including an

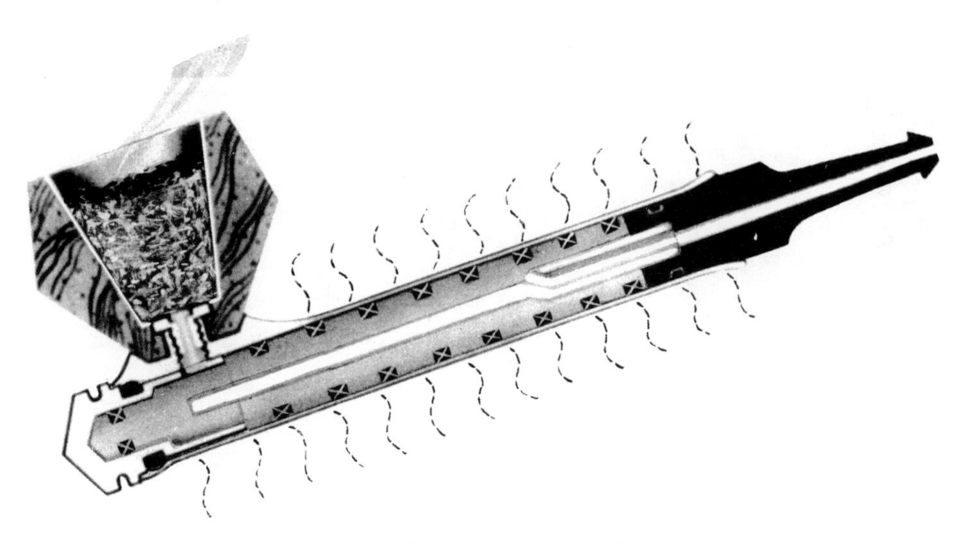

aluminum stem that cooled the smoke, a tissue-paper filter that cleared impurities, and a bowl that burned more efficiently. The result was so superior a smoking instrument that even his doctor asked for one. Kirsten patented his design the same year, and asked a young neighbor, Chauncey Beach, to manufacture them in Kirsten's home shop.

Kirsten's son Gene recalls: "People slowly got wind of it, and pretty soon my mother was spending all day going to the door, making change and selling pipes." Kirsten began selling the pipes through a chain of shops called the Brewster Cigar Stores, and they became so popular that he had to make a decision whether to quit his teaching position or concentrate on pipe sales full-time. He chose to stay at the university, and sold the manufacturing rights for the pipe to a man named George Gunn in 1938. Gunn built the business to the point where the Kirsten Pipe accounted for 40% of the total pipe sales in the United States at the end of World War II. He encountered financial difficulties soon after, however, and in 1949 the Kirsten family bought the business back from him. The Kirsten Pipe Company, now run by Gene Kirsten, still does a brisk trade, selling several thousand pipes yearly to clients around the world, primarily through mail

Smokers are still puffing away on Professor Kirsten's smooth-smoking metal pipe.

order. (The company's main output now is small, precision-made parts for electronics gear, diesel engines, and assemblies that go into "just about every section" of Boeing airplanes.)

None of Kirsten's many other commercially oriented inventions met with the financial success of his pipe, but most did reasonably well and all, his son says, were "greeted with respect." Kirsten invented a fire extinguisher that used specially treated dust, one of the first uses of chemicals to stop fires. He also devised a vortex "air-washing" machine that cleaned timber mills and plastics plants of "wood flour," as well as neon signs that operated on ordinary 110/220 current. (Examples of the latter once adorned the exteriors of Seattle's Fifth Avenue Theater and Edmund Meany Hotel.) The professor was inspired to create a violet-light "death ray," designed to eliminate the codling moth pest from eastern Washington apple orchards, when he noticed that a wasp was killed by the high-intensity electricity of one of his neon signs. During World War II, he devised a parabolic air-raid siren so powerful that it could be heard 30 miles away, but so directional that when pointed away from the listener it was virtually inaudible.

One Kirsten invention that never went beyond the design stage was the Utopian Bed, designed on a lark after a sleepless night in the late 1930s. The plan called for a fabric air mattress kept taut by a compressed-air tank in the basement; softness and temperature would be adjustable, and the fabric would be porous enough to allow heated air to slowly seep out during the night — thus eliminating the need for blankets. According to one story, another of Kirsten's creations grew out of professorial impatience. Frustrated with the amount of time it took to dress in the morning, Kirsten allegedly devised a jumpsuit that incorporated trousers, shirt, vest, and jacket in a one-piece outfit. The whole affair zipped up smoothly, and the necktie was then attached to the zipper pull. (Gene Kirsten doesn't recall such a gadget, but says that "it sounds like my father.")

But the invention Kirsten thought was his most important, the one he hoped would revolutionize both air and sea travel, was never widely accepted. The cycloidal propeller, designed for both airplanes and ships, used a simple principle: instead of the drive shaft reversing rotation to allow the vehicle to change direction, as with a standard engine, the drive shaft turned in only one direction, but the angles of the blades changed. Even big, clumsy boats could turn in small areas, dock sideways, or make abrupt stops at high speeds. The *Seattle Post-Intelligencer* reported in 1925 that six Prohibition-era cops took a spin in an experimental model and were suitably impressed. According to the paper, "If the government ever adopts [this] revolutionary boat design, the price of Scotch whiskey is due to soar far past the reach of ordinary mortals."

The Boeing Company and Kirsten went into partnership to produce the propellers for use on dirigibles and airplanes, but the idea fell from favor, mostly due to its complexity and cost. Nor did it catch on for use in ship propulsion, except in a limited way. The German firm Voith-Schneider bought the manufacturing rights to the prop and began making a simplified version of it in the 1930s. The design is still used in Europe and on a few American tugs. (Foss Boats in Seattle, for example, has six Voith-Schneider cycloidal tugs.) Kirsten never lost faith in his idea, however, and at the time of his death in 1952 was at work on a system of cycloidal windmills for power production.

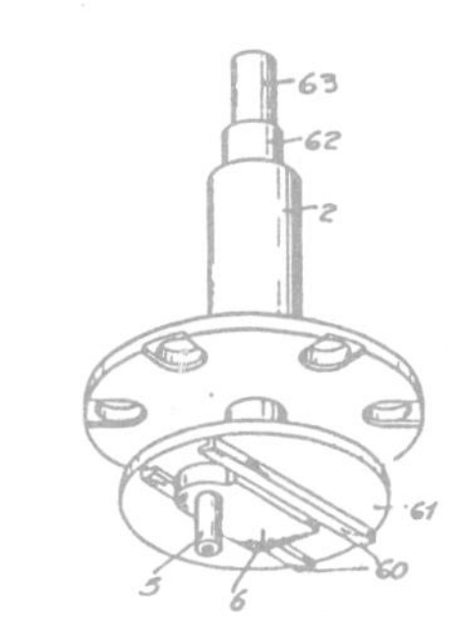

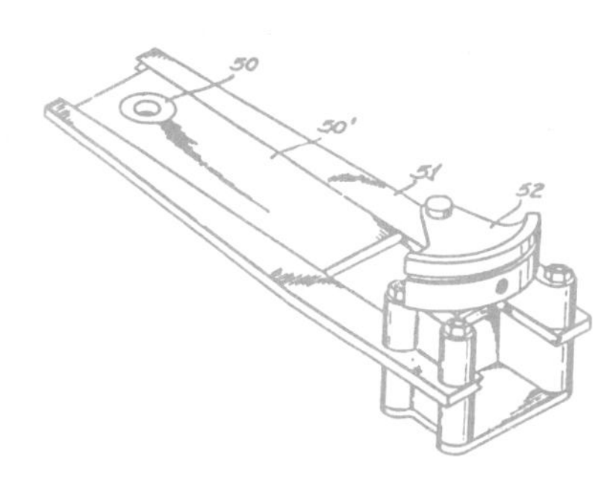

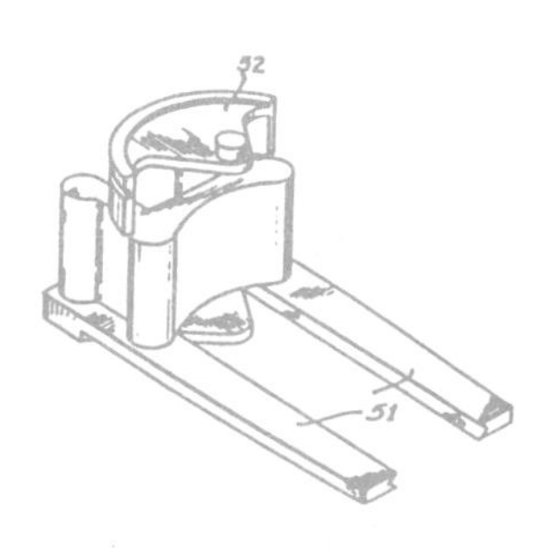

Details of the patent drawings for the control mechanism of the cycloidal propeller, which Kirsten hoped would revolutionize air and sea travel.

The Kwik-Lok

1952

Handy little items, aren't they?

The Kwik-Lok is an unassuming little plastic tab used by millions of people around the world every day to close bread bags and other types of plastic bags. It is arguably the most practical invention ever created in the Northwest, and almost certainly the most common; an estimated four billion are used annually by U.S. baking companies alone. The invention and proliferation of the Kwik-Lok occurred at about the same time that plastic superseded wax paper as the usual wrapper for bread — a practice that also began in the Northwest.

The Kwik-Lok was invented in 1952 by Floyd Paxton of Yakima, the president of a small company that manufactured automated equipment for nailing wooden packing crates for fruit. He needed a new product to manufacture, since cardboard boxes were coming into use as an alternative to wood crates. A Yakima produce house, Pacific Fruit and Produce, had recently started using plastic bags for storing apples; they closed them with ordinary rubber bands. Pacific Fruit's owner remarked one day that if Paxton could come up with a better "closure," as the devices are known in the industry, he'd be very interested.

Paxton whittled a prototype closure out of a flat piece of Plexiglas. It was much superior to other closures for plastic bags then in use: wire twists broke easily and were clumsy, and sticky-back tape was unreliable and not always reusable. He named his prototype the Kwik-Lok and designed a hand-operated machine that punched them out one at a time. When Paxton showed the closure to the head of Pacific

Fruit, he told Paxton to make a million of them immediately. Paxton then designed a die-punching machine that could produce the closures at high speed.

Paxton tried several times to patent the Kwik-Lok, but his appeals were rejected because the design was held to be similar to a package-and-closure combination patented by an Englishman named Brownfield in 1934, and a system of hooks and enclosures for "shaping and supporting meat" patented in 1941 by an American named Wohlmuth. Paxton did succeed, however, in patenting machinery to automatically attach large quantities of Kwik-Loks — already imprinted for easy color-coding, labeling, and pricing — to the "pony tails" of plastic bags. The first of these machines attached closures individually from a gravity feed, but they were later modified to incorporate continuous strips of clips (called Strip-Loks) that were separated as they were attached to bag ends. The ability to attach closures rapidly to large numbers of bags led to the Kwik-Lok's popularity among food producers.

Paxton grew wealthy as a result of his product, and spent much of the money to further his ultraconservative political views via newspapers and other media. During the 1960s he was national president of the John Birch Society, he ran unsuccessfully as a congressional candidate on four occasions, and he published the right-wing *Yakima Eagle*. He also continued to dabble in inventing; in 1968 *The Seattle Times* reported that the then-king of Kuwait had invested a million dollars in a Paxton venture (ultimately unsuccessful) to grow barley hydroponically for cattle feed.

Poly-Bagging Machinery for Bread Wrapping

1965

For years, bread arrived in stores and was sold to the public in plain wax-paper wrappers. But in 1955, a salesman named Larry Smith told Dean Brown, then manager of the in-house bakery division of Albertson's Food Stores in Boise, Idaho, that "several retailers in the East" were using a polyethylene bag to wrap their breads. They found that plastic bags kept bread moist and fresh for longer periods of time than paper wrappers, Smith said. Albertson's tested the idea and the company's bread sales increased dramatically. It was not until some time later that Smith admitted to Brown that he had been bluffing, that in fact the use of plastic bags was his own idea.

The idea caught on quickly among Northwest bakers. Probably the first use of poly bags for wholesale bread sales was a hand-bagging operation set up in 1957 by R. D. Hoyt of the Hansen Baking Company in Bellingham. (LeConie Stiles Jr. of Ashbrook Bakery in Seattle began a similar operation only a few weeks after Hoyt.) Hoyt is quoted in the trade journal *Baking Industry* in August 1966: "By golly, that first year our sales tripled. We sold an awful lot of buns." Not all bakers were keen on the new method — most had considerable amounts of money tied up in old-style wrapping machines — but the practice began to spread, and soon a strong need developed for automated bagging equipment.

Jere Irwin, Floyd Paxton's son-in-law, together with Paxton and another Yakima engineer named Ted Markquist, designed the Mark 50 automated bread-bagger in 1962, probably the first completely automatic bread-bagging machine and still the prototype for the various types of bread-baggers used today. The first unit went to Snyder's Bakery in Yakima and was designed to handle bags of various sizes. The bags were hung on wickets by holes in their lips, in batches of 500 or 750; each bag was blown open, pulled off the wicket, and pulled over a loaf of bread. The manufacturing rights to the Mark 50 were sold in 1965 to the AMF Corporation, which continues to produce it.

Superefficient Wood Stove

1985

Americans' interest in renewable resources peaked in the late 1970s and early 1980s, following the global oil shortage, when large amounts of federal and local money were allocated for studies of solar, wind, and biomass (wood and wood-related) energy. Entrepreneurs introduced such products as solar panels, high-tech windmills, and improved wood stoves, and homeowners took advantage of tax breaks if they invested in energy-saving programs. During the 1980s, however, the Reagan administration, seeking to protect the fossil-fuel (read: oil) industries, eliminated the tax advantages and cut back funds. Stringent EPA testing, meanwhile, virtually stopped independent development.

Recently the tide has turned, and public interest in renewable resources is again stirring. In the Northwest, this interest is spurred by evidence that our surplus of energy — a luxury the area has always enjoyed — is slipping away. Electricity trunk lines are operating at peak capacity. Public agencies are discussing ways of coping with brownouts for areas that have never before considered the possibility. Even the relatively mild 1989–90 Puget Sound winter caused Seattle City Light to move into emergency preparedness, ready to ask industrial customers to shut down.

During the peak period of research on renewables, one area that was seen as providing at least a partial solution was biomass: trees, grasses, bushes, seaweed, and other flora. (Biomass waste in the Northwest is mostly logging and agricultural residue and municipal solid waste.) But home wood stoves, the usual means for generating biomass energy, never offered stiff competition to relatively cheap and efficient oil and gas burners. Their pollution levels were, and are, still high, their efficiency levels are lower than oil and gas burners, they can be unsafe if improperly installed, and cordwood is cumbersome.

For nearly 20 years, Larry Dobson has addressed the challenge of making wood stoves a viable source of energy. Aided by grants from the Washington State Energy Office, Vaagen Timber Products of Colville, Washington, and the U.S. Department of Energy's Office of Energy-Related Inventions, he has been working to perfect a stove that will burn, with virtually no pollution or heat loss, all types of wood waste — sawdust, chips, pellets, slash, or the logging residue known as hog fuel. A patent for his design was granted in 1985. By using only waste and by improving pollution levels, Dobson hopes his stoves will someday make it economically and environmentally feasible to use wood energy to heat greenhouses and factories as well as homes: "The wood waste generated in logging is just phenomenal. Mills have a terrible problem disposing of it. They spend hundreds of thousands of dollars a day on it. With even just a tiny part of the residue from logging, we could keep every

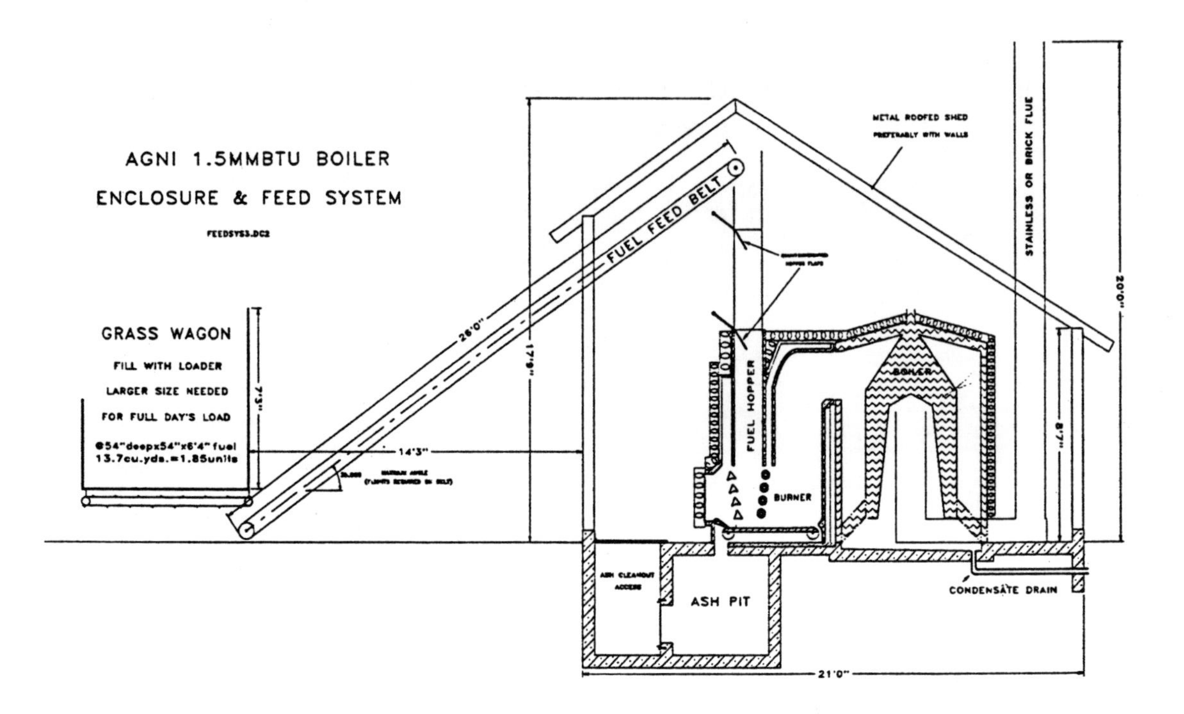

Above: This huge wood-burning furnace will be installed in a 1,000-square-foot production facility being built by Pyro Industries of Burlington, Washington, a manufacturer of pellet stoves.

house in the region heated year-round."

Information from the Northwest Power Planning Council, the interstate agency charged with developing regional energy policy, backs up Dobson's claim. Half of Washington's timber-related biomass waste, they report, is currently destroyed through prescribed burns; the rest goes into pellets, pulp, and plywood. According to a paper the council prepared in 1989, the energy potential of the state's available biomass waste is 190 trillion BTUs per year, far more than the amount required to serve all the energy needs of all the houses in the state.

Dobson approaches his wood-stove research from an eclectic background, which includes iron forging, home design, solar energy, carpentry, and fine woodworking. "I've always been interested in figuring out ways of doing things. I've also always been fascinated with fire and pyrotechnics. As a kid

I had a lab in my basement, and I made a lot of bombs and rockets." Born in 1941 and raised in Lake Forest Park, north of Seattle, Dobson majored in chemistry at Antioch University in Yellow Springs, Ohio, and worked for Monsanto Chemical as an analyst, but "got disillusioned" and switched to the study of political science. He spent several years working in Germany and California, and a Peace Corps job took him to the Uttar Pradesh state of India, where he helped build a biogas-energy production plant.

Back in the Northwest, inspired by a variety of sources (including Buckminster Fuller's design theories, Peter Pierce's book *Structure in Nature as a Strategy for Design*, the appropriate-technology movement, and Eastern philosophy), Dobson designed and built everything from small projects — highly efficient hand tools, huge stilts, multiconfigurable playground equipment, educational toys based on the fundamental principles of geometry — to larger projects such as greenhouse solariums, vertical-axis windmills, ferroconcrete hot tubs, and entire homes. He designed a house on Whidbey Island based on the principles of crystal geometry; another sported a roof shaped like a huge fish (rainwater drained down the fins). None of these projects was marketed commercially, however, and they became secondary once Dobson began investigating biomass.

"I got started on the [wood stove] project because I got tired of having to cut firewood and tend to the stove. There was all this free sawdust at the mill, so I designed a burner that could use that and make it last for a long time. Things just sort of went on from there." Dobson analyzed the main problems with conventional wood stoves as follows: inadequate handling of the fuel's moisture content, uneven carburization, and temperatures too low for adiabatic combustion (that is, combustion without heat gain or loss). His

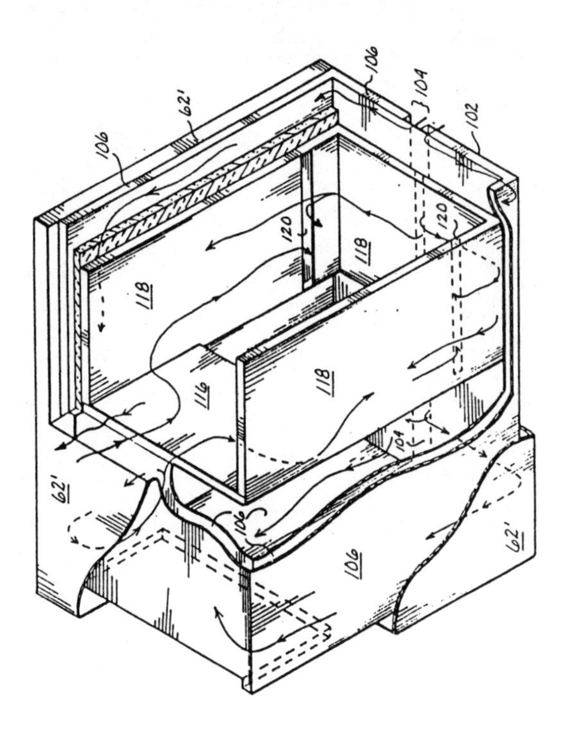

Larry Dobson's superefficient wood stove applies high technology and environmental concerns to an ancient means of keeping warm.

design aims at optimum efficiency through drying the wood thoroughly before burning it, then retaining heat in the combustion chamber until the wood is completely combusted. This prevents the gases from burning with too little air, which makes for sooty and incomplete combustion, or at such hot temperatures that the fuel is consumed too quickly.

Dobson's basic design has been developed piecemeal over the years. His home model, installed in his Seattle basement, is roughly the size of a standard oil furnace: 4'6" × 2'6" × 4'. The main body is squared off, with the guts hidden by a 20-gauge mild-steel casing. At the top front is the stainless-steel hopper; the bottom front has a heavy door that allows access to the combustion chamber. Emerging from the center top of the furnace is the stainless-steel plenum, the flue through which hot air is funneled into the house. A small rank of programmable electronic controls connects to inlets on the side panel. Small windows on the side of the furnace afford views of the combustion chamber, made of

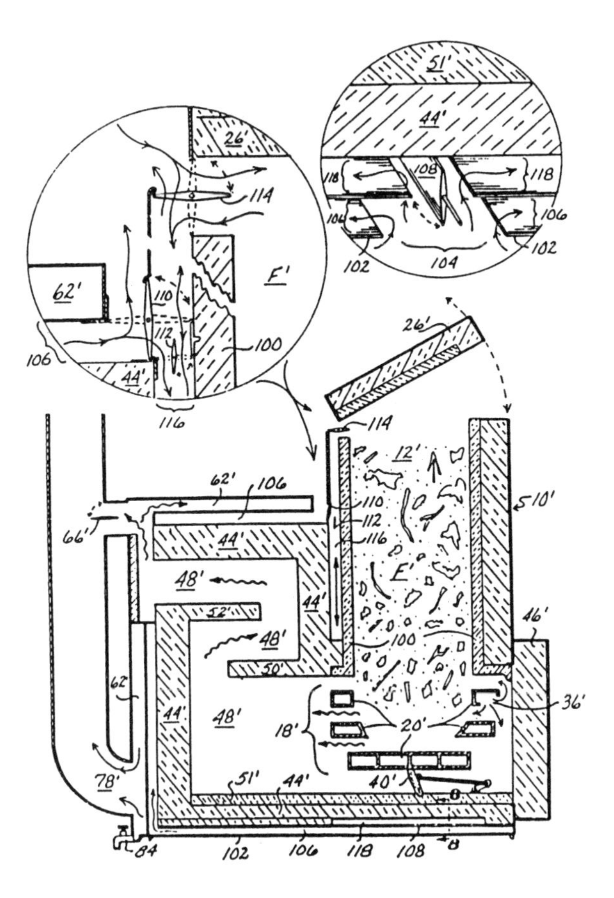

Turned up high, Dobson's home stove model will put out 150,000 BTUs of heat per hour, far more than the average house would ever use, with virtually no pollution or heat loss.

nonporous ceramics capable of maintaining long-term temperatures that far exceed those at which aluminum melts. A change of the controls, altering the ratio of air to fuel, makes the fire in the combustion chamber glow bright orange, then deep red, then bright orange again.

Fuel is held in an automatic gravity-feed hopper with a capacity of about three cubic feet of sawdust or wood chips. A pickup-truck load will heat Dobson's large, drafty house for one to three weeks. With the stove turned low, a single hopper load lasts overnight. During the day, Dobson and his housemates typically stoke it once every six hours. (This limited hopper capacity is only one of the problems that Dobson has yet to overcome. "Bringing the fuel to the house is a hassle. But the average homeowner has access to the street. If wood waste becomes a widespread fuel source, it could just be blown in from a truck on a weekly basis," like oil deliveries.)

Dobson's stoves have higher efficiency quotas and cleaner performance than any wood stove currently available. Commercial stoves average a rating of 8 on the Bacharach 0–9 scale of smoke-spot density (5 is "very poor"); Dobson's test near 0. Mark Hooper, an environmental engineer for the Seattle office of the EPA and a former member of its Biomass Utilization Task Force, says: "Ideally, what you want with a wood stove is to convert all your fuel into CO_2 and water, because then you'd have one hundred percent efficiency overall. ["Overall" refers to the product of the heat transfer efficiency times the combustion efficiency.] Most wood-burning systems get nowhere near that, and a lot are lucky if they get fifty percent. Even the best, the so-called 'airtight' models, run about seventy to seventy-five percent. Larry's designs test out with an overall efficiency of about ninety-five percent."

Dobson spends much of his time these days promoting an industrial-size version of this stove. The new project, still in the design stage, is huge: a 9' × 14' low-pressure hot water/steam boiler. A 630-cubic-foot hopper will provide it with 24 hours' worth of green chips, bark, or sawdust — 150 tons of waste per month. It will produce 1,500,000 BTUs/hour (440 KW), with 99.99% combustion efficiency and 80% heat transfer efficiency. Dobson estimates that it will be cheaper than an equivalent oil or gas burner, with monthly savings of $4,500 to $5,200 over natural gas, $8,900 to $9,640 over electricity, and annual savings of $30,000 to $100,000. He recently arranged to install the new furnace in a 1,000-square-foot production facility being built by Pyro Industries of Burlington, Washington, a manufacturer of pellet stoves.

The Self-Cleaning House

1950s

Rooms that wash themselves down at the touch of a switch; clothes that are laundered and dried as they hang in the closet; dishes that are automatically scrubbed right in the cupboard. Fireplaces that never need manual emptying; organic waterless toilets; tubs and sinks that never need scrubbing. Welcome to Frances Gabe's lifework: the self-cleaning house.

Gabe (a pseudonym, taken because her family is "embarrassed by what I do") has spent over 30 years perfecting every nook and cranny of her remarkable dwelling. Her mission, she says, is to free housewives (and househusbands) from boring, everyday chores: "All the labor-saving devices that have been invented so far aren't really labor savers. Housewives have always had to be on their knees, or with their heads in a hole somewhere. The dishwasher, the clothes washer, the dryer, the tub, the whole works — I just had to do something about it."

The prototype for Gabe's self-cleaning house is her own home in Newberg, Oregon, built largely by herself. (She is the daughter of a contractor-architect and the former wife of another contractor.) Many of the materials she used, such as the marble floors, timbered ceilings, and windows, are second-hand or scavenged. She estimates that the whole house — living room, dining room, kitchen, dressing room, bedroom, bathroom, and hall — could be replicated, complete with labor-saving devices, for about $18,000. "I've been called a liar for saying that, but it's true. If you used all new materials, figure on double that cost, at least, and if you hired someone to do it, it'd cost about the same as an ordinary

Above: The 1950s artist's conception above (not Frances Gabe's, but similar in spirit) shows how the Self-Cleaning House could make quite a splash.

house. I'm not interested in making a lot of money from this thing. What the heck? I'm seventy-five years old, how long do I need to hang on to a million dollars? I just believe very strongly in the home. I think that when the home goes, America goes, so I'm doing my best to make everything as inexpensive as I can. That way, anybody could afford one."

On entering Gabe's house (she designed the back door especially for short people like herself), visitors can view over 60 separate inventions. Every room is equipped with a special cleaning unit in the ceiling; nozzles direct a moderately pressurized warm water/detergent spray around the entire room, while slightly sloped floors ensure proper drainage. Rooms are dried after cleaning by forced air blown through the same pressurized system. Bookcases, furniture, stereo gear, pictures, and other objects stay dry without being covered up because they are "especially designed and made to reject moisture." Gabe's organic, waterless toilet has a disposable liner that slides over the hole as the lid lifts ("so you don't have to look down the pit, which I think is a *very* desirable thing"); the liner, which descends along with the refuse, is pretreated with biodegradable material such as vegetable fibers or soda to speed decomposition. The "clothes freshener," sectioned off in one part of the bedroom closet, washes and dries clothes automatically as they hang, in about the same amount of time as does a regular washer/dryer. Dishes are cleaned in a similar fashion as they rest in the kitchen cupboards. Sinks are equipped with foot-level valves with rounded tops; the user steps on the valve and rolls it one way for cold, the other for hot. This feature, Gabe notes, would be especially useful for the physically disabled or frail.

Gabe says there are only two housecleaning problems she hasn't yet solved. One is the question of carpeting; removable area rugs work well in the self-cleaning home, but wall-to-wall carpeting would stay damp. "Who wants to walk on wet floors? *I* don't. Though you never know...in some hot countries it might be all right. It might feel pretty good, in fact, and, besides, it'd dry quickly." The other problem that Gabe hasn't yet resolved concerns personal neatness: "I haven't invented a way to pick up after people. I can't put a motor into anybody's backbone."

The long process of testing her inventions is nearly over, Gabe says, and the self-cleaning house is finally ready for the market. "I'm extremely happy these days, because I've whipped the whole package into shape, and it really honest-to-God works. Now I want to let the world know. I'm doing my level best to get it on the market, and my patents are all applied for — although they're having a heck of a time finding a pigeonhole to stick me into." Gabe is more interested in presenting the self-cleaning house as a totality than in marketing her inventions singly, although she notes: "I have had to change my ideas about this. Many people with existing houses would still want to remodel to accommodate the self-cleaning house in their baths and kitchens. You don't get everything just because you want it." She is currently producing a series of portable, suitcase-size scale models of the house as part of her campaign to interest manufacturers, although so far she has had little success in finding backers: "Every time somebody gets enthusiastic about it, they get scared off later when they realize the enormity of the project."

FATHER'S DAY

Father's Day, always a red-letter day for school-age children, was created in the early part of this century by Sonora Smart Dodd (Mrs. John Bruce Dodd) of Spokane. Inspired by her own father and by a Mother's Day sermon at her church in 1909, she came up with the idea of a day honoring fathers and worked unceasingly to make its observance widespread. By 1914, President Woodrow Wilson had acknowledged it, and it became a nationwide (if unofficial) observance. In 1937, Congressman Charles H. Leavy said on the House floor that "through the efforts, energy, and tireless activity of Mrs. John Bruce Dodd, a constituent of mine, of Spokane...officials recognized the significance and importance of having a day on which, throughout the Nation, tribute would be paid to the fathers of America. Now, by common consent, the third Sunday in June is recognized as Father's Day."

Goose Valve

1989

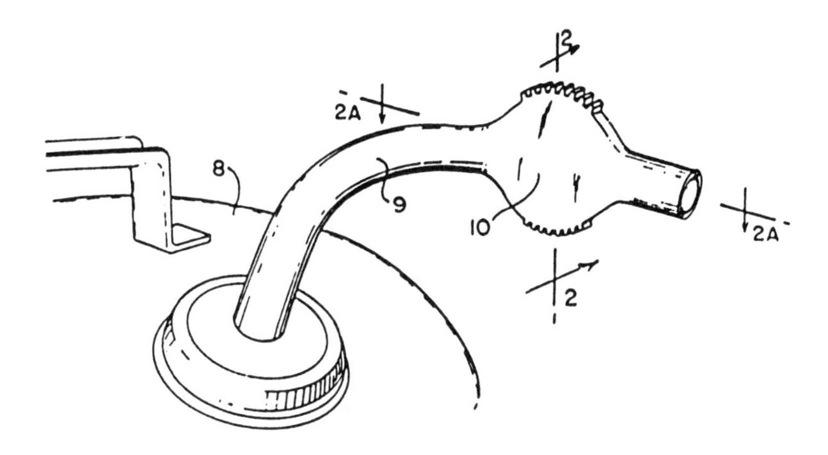

Although he initially designed it to aid in pouring gasoline, Dan Vorhis hopes his Goose Valve will be handy in situations such as dispensing drinks in cafeterias.

Dan Vorhis is a product designer for Mountain Safety Research in Seattle, which specializes in high-quality outdoor gear. (He created a popular foldable canteen for the company and helped design their highly efficient portable water filter; he is also MSR's liaison with aspiring inventors who contact the company.) But Vorhis is also an independent inventor and is representative of the many hopefuls who do their own research and development, write patent applications, and market their products themselves. His current project is a valve he invented to provide fingertip control for gasoline cans and other pourable containers. He informally calls it the Goose Valve, a name suggested by his sister (from to the verb "to goose," the expression "loose as a goose," and the raucous duck-call sound it makes if the inventor blows hard through it).

Vorhis says: "I was working as a landscape gardener, and I got tired of spilling gas on my feet every time I'd try to fill up a lawn mower or weed-eater — so I decided to do

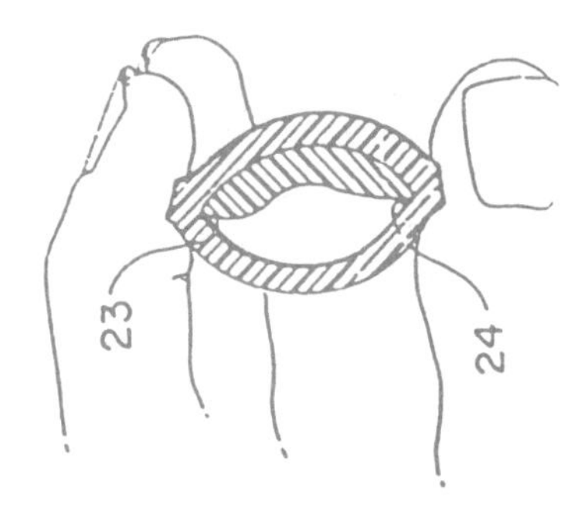

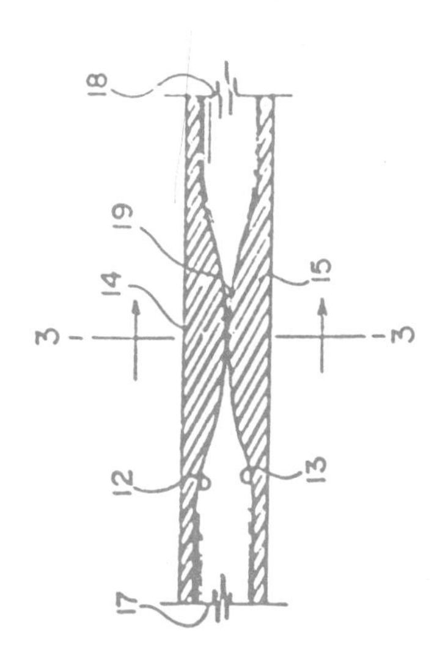

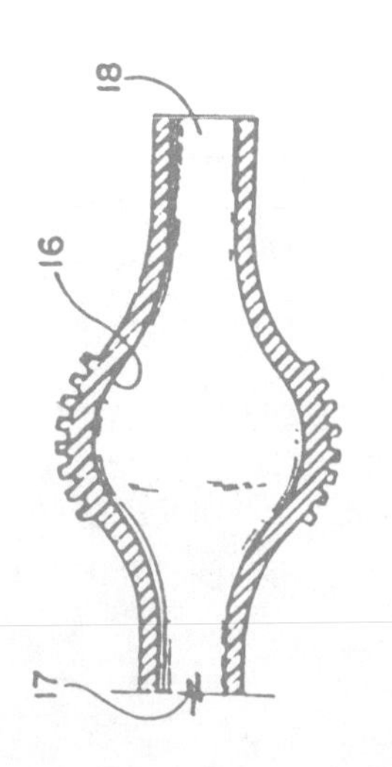

something about it." The valve he created consists of two thin plastic wafers (placed flat, end against end, they resemble a flying saucer). When set crosswise into a tube made from a pliable plastic called an elastomer, the wafers form a natural seal, but when lightly pressed on the edges with the fingertips they open and let liquid through. It is inexpensive (Vorhis figures the retail price will be two to three and a half dollars); fluid cannot be contaminated by material from the outside, since it is completely sealed; it has no wearing, toxic, or rustable parts; and the hydraulics are elegant, with little turbulence disturbing the flow.

Vorhis has kept a file of potential ideas for years, including the Goose Valve, but decided to concentrate on it in the mid-1980s, shortly after moving to Seattle, because "the timing was just right." The catalyst came when Vorhis returned to its owner a wallet he had found on the street. The grateful recipient was a plastics engineer who let Vorhis pick his brain about injection molding for plastics. He also introduced Vorhis to the Inventors Association of Washington, a now-defunct organization of aspiring inventors who met regularly to exchange advice, support, and information. Vorhis eventually edited the group's newsletter, an experience he says helped him "find out how the whole system works." He quit his job and worked full-time for two years on the development of a prototype for the valve, informally apprenticing himself to a plastics moldmaker while his girlfriend (now his wife) supported him.

Vorhis was granted a patent for the valve in 1989 and is currently raising money to start a production company. He is sanguine about his prospects: "Even if it doesn't work, or if I get ripped off, I feel like I'm even on this one — I've learned so much about how this stuff works. Besides, all I want is enough to buy my farm."

The Art of Medicine

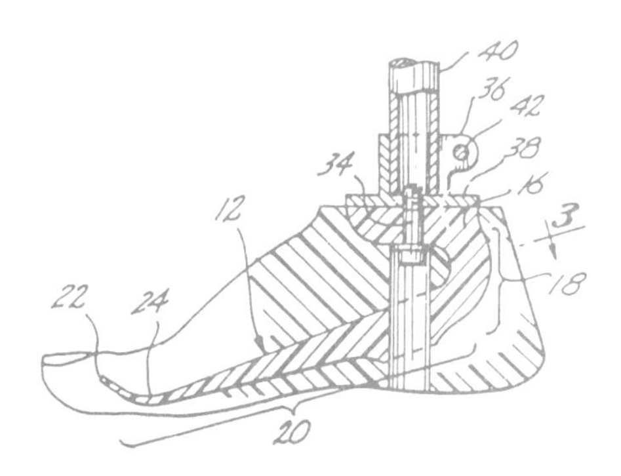

Heart Defibrillator

1959–62

Perhaps the most dramatic medical innovation to emerge from the Northwest is the heart defibrillator, developed between 1959 and 1962 by Dr. Karl William Edmark, a cardiovascular surgeon in Seattle. A later modification, a lightweight, portable defibrillator developed by a team at Edmark's company, Physio-Control, was a key ingredient in the formation of Medic One, a pioneering mobile rescue unit launched in Seattle in 1970.

The defibrillator counteracts ventricular fibrillation, a form of heart disease in which the muscles of the ventricles, the lower two chambers of the heart, twitch spasmodically. The heart normally controls the pumping cycle of its four chambers by means of electrical signals that begin in the organ's natural "pacemaker," a small bundle of cells in the right atrium called the sinoatrial node. These signals cause the atria, the heart's upper two chambers, to contract in sync with the ventricles, the lower chambers. When the heart fibrillates, this synchronism is disrupted and it is unable to continue pumping blood. Unless the blood flow can be maintained through cardiopulmonary resuscitation (CPR), the patient will die within three to four minutes. By placing the defibrillator's pair of paddle-like contacts on the patient's chest, a doctor or paramedic can apply an electrical shock that jump-starts the heart.

The defibrillator is not Edmark's only invention; he holds about 10 patents for various biomedical products, the first of which was a simple but effective heartbeat indicator. More sophisticated versions of Edmark's monitor are now

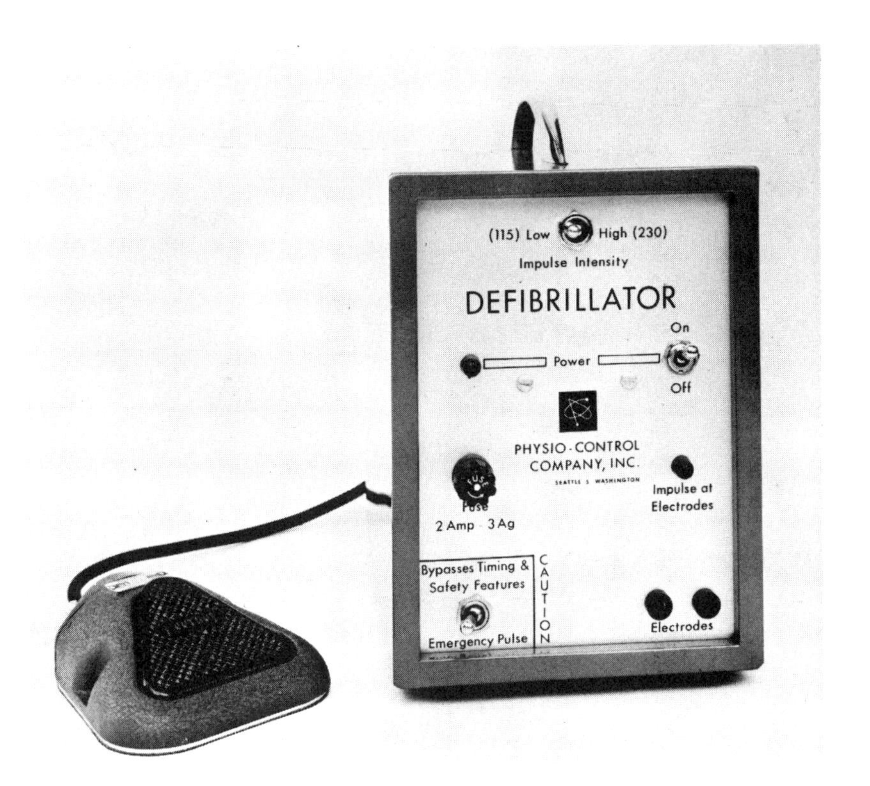

An early AC defibrillator, one of the few manufactured by Physio-Control before it switched to the more reliable DC configuration.

standard equipment in hospitals and clinics, but in the early 1950s, when he began his work, they were virtually unknown. Edmark, a former ham radio and electronics enthusiast, was finishing his surgical residency at the Lahey Clinic in Boston when, on card tables set up in a spare bedroom, he began work on an indicator that would flash a light each time the heart pulsed and would sound an audio alarm if the beat faltered. (Undetected cardiac arrest was the principal cause of death in the operating room, since doctors were unable to determine the precise moment when a patient's heart failed. By the time such failure was detected it was often too late.) Edmark patented his heartbeat monitor in 1955 and formed Physio-Control to manufacture it.

By the late 1950s, after his move to Seattle and association with the cardiovascular teams at both Swedish and Providence hospitals, he became aware of the need for another instrument, one that could reliably reverse the effects of ventricular fibrillation. About 90% of all heart failure was attributable to fibrillation, and the risk was even higher during heart surgery. Techniques for closed-chest resuscitation, such as CPR, had not yet been fully developed; to save a fibrillating patient, a surgeon had to open the patient's chest and directly massage the heart, or use a relatively crude alternating current (AC) defibrillator, which delivered electrical charges to stop and restart the heart. AC defibrillators, which had been used experimentally since 1948, were clumsy and unreliable; they plugged directly into the wall and used ordinary electrical current, with a voltage too high for safe operation. Often the power requirements of a defibrillator plunged the entire hospital into darkness.

In 1959, under the auspices of the University of Washington and the National Institutes of Health, Edmark experimented with a direct current (DC) machine that would deliver a charge at a much lower voltage and in a much shorter time span: 1 or 2 milliseconds, as opposed to the 250 milliseconds previously used. He recalls: "I had read a paper by a guy named McKay, who had done some work in using capacitors in defibrillating cats. He reported that his success rate went up if he included a one-henry inductor in series in his setup. [A capacitor is a nonconductor, such as a battery, that can hold and then discharge electricity. An induction coil creates high voltage by sending direct current through a primary coil to charge a second coil; inductance is measured in henrys.] I had a little lab down at Pacific Northwest Research, the predecessor to the Fred Hutchinson Research Institute, and I set up what McKay had

An early DC defibrillator.

described — a capacitor in series with an animal.

"I tried it just once. We got the most horrible, titanic muscular response, and the poor animal broke all the restraining straps. It was just awful, and we never tried it again. But I had an oscilloscope hooked up, and out of the corner of my eye I thought I saw a very sharp, high-voltage transient that nobody else saw. I thought about this afterward and I wondered: do you suppose the conductors in that transformer were saturated? If you make a transformer, and you put so much current through it that it saturates the metal inductor, it then loses its magnetic properties and you have, in effect, a coil of wire. I took the transformer apart, took the coil of wire out, then took another one apart and put them together so that they were mutually reinforced. When I put them together, it turned out to be only about thirty millihenrys — that is, .030 henrys. The iron core in the inductor was one henry, and that turned out to be about right. So then I went back and put the coils into the thing, charged it up to about three thousand volts, hit the foot-switcher we were using at the time — and nothing happened. Then I looked over at the monitor oscilloscope and I saw the animal was defibrillated.

"I went back and did it again, and I realized there was a very slight little 'ping.' It turned out that a thirty-two-millihenry inductor and a capacitor with about thirty-two microfarads was just the right combination to give a critically dampened one-millisecond waveform across the chest. [A farad is a measure of capacitance.] So that was that; if I hadn't noticed that one transient, and if I hadn't wondered whether the inductor might have saturated, I wouldn't have invented the defibrillator."

Edmark continued testing his defibrillator on dogs, a practice about which he now has some reservations: "For years now, I have not done anything related with biological experimentation. I got completely sick of it. I'm sorry to say it was necessary to destroy a lot of test animals to get that defibrillator working." In 1961, the Edmark defibrillator was put on standby during open-heart surgery on a 12-year-old girl; when her heart fibrillated midway through the operation, the instrument successfully defibrillated her and she recovered.

Edmark did not patent the defibrillator, feeling that it should stay in the public domain. Instead, Physio-Control concentrated on becoming the dominant manufacturer of defibrillators and related equipment, a position it continues to hold.

MacGregor Rejuvenator

1933

On display in the Museum of Questionable Medical Devices in Minneapolis, Minnesota, is the MacGregor Rejuvenator, one of the countless quack medical devices and "snake oil" remedies presented to a gullible public over the years. In 1933, William E. Mortrude of Seattle developed the "MacGregor Method of Rejuvenation," purported to help reverse the aging process. Treatment consisted of bombarding the patient/victim with radio waves, magnetic fields, and infrared and ultraviolet rays while he or she lay in a 500-pound "electrical treatment cabinet" that resembled an iron lung. Only five of these machines were manufactured.

Kidney Dialysis Equipment and Techniques

1960

The University of Washington, in the late 1950s and early 1960s, was fertile ground for the development of new forms of medical technology. One of the most important breakthroughs came when a team of researchers led by Dr. Belding Scribner, a nephrologist, Dr. Albert Babb, a professor of nuclear engineering, and Wayne Quinton, a biomedical technician and instrument builder developed the world's first practical, ongoing treatment for kidney failure. Their pioneering techniques and hardware led directly to the formation of the world's first kidney center, at Swedish Hospital in Seattle, and to safe, affordable treatment for hundreds of thousands of patients worldwide.

Uremia, the medical term for renal failure, occurs when the kidneys can no longer carry out their normal function of cleansing and purifying blood. Prior to the 1960s, uremia meant quick and certain death for tens of thousands of patients annually. Various solutions had been tried over the years: kidney transplants were performed successfully in some cases, but they were still highly experimental; even today they remain problematic and uncertain. Another method was to use a machine to cleanse the blood, a process called dialysis. A Dutch physician named Dr. Willem Kolff created a dialysis machine in 1939, but it was experimental and impractical. Improvements were made in Kolff's machine later, but it remained cumbersome and far from foolproof.

A major problem with dialysis was that patients' arms or legs had to be hooked up to machines several times a week. Fragile arteries and veins could only take so much

The dialysis machine perfected at the University of Washington paved the way for reliable treatment of kidney failure.

abuse before they suffered complete collapse. The challenge was to find a way to connect patients to machines while keeping the blood vessels from weakening. In 1960, Scribner, a Stanford-trained surgeon who had joined the UW staff in 1951, devised a shunt — later known as the Scribner Cannula — that diverted the flow of blood to a U-shaped tube permanently installed along the outside of the patient's arm. During normal activity, the tubing would simply be an extension

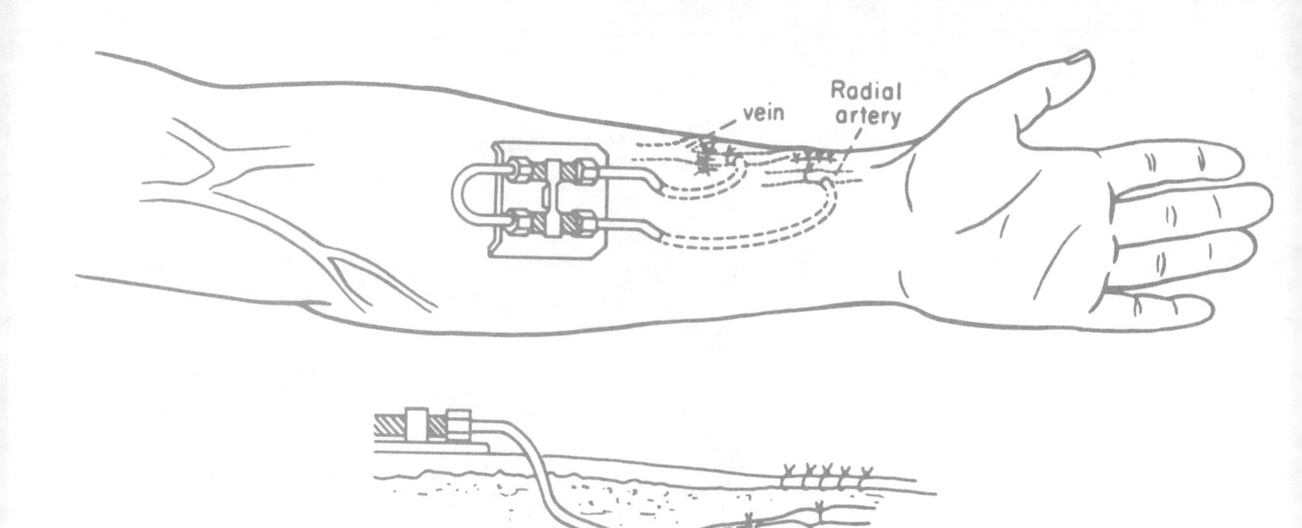

With the Scribner Cannula, kidney patients could connect with dialysis machines but avoid vein collapse.

of the circulatory system. When the shunt was removed, the patient could be connected to a dialysis machine without the necessity of new incisions.

Scribner enlisted a team of colleagues, including Babb and Quinton, to help him work out the details of the device. Dr. Loren Winterscheid suggested the use of Teflon for the shunt's tubing material because it did not react with human tissue; only later did the team discover that the material's nonstick property was another important ingredient in its success. They developed a metal plate on which the shunt could rest flat against the arm, and used plumbing equipment to fashion the connections.

Later that year, the first cannula was inserted in the arm of a Boeing machinist, Clyde Shields. This was done without prior testing of the technique on animals, a process that could not occur under the more stringent medical guidelines used today. The shunt, though in many respects a great success, still had serious problems. For one thing, Teflon was stiff and irritated the skin around the punctures, causing inflammation. The shunts tended to clog over time, and only lasted three or four months before they needed replacement. By early 1961, Scribner and his colleagues were running out of sites in which to insert new shunts in Shields. But the problems were gradually worked out. Quinton overcame the shunt's stiffness by developing a "shock absorber" that was extremely pliable and could remain implanted for long periods. The plumbing equipment and other cumbersome materials were likewise replaced by more sophisticated hardware.

After perfecting the shunt to their satisfaction, the team, armed with a grant from the Hartford Foundation, turned its attention to creating a more efficient dialysis machine. The units then in use weighed about half a ton, including the bulky dialysis fluid; they were also extremely expensive and could only service one patient at a time. Babb and his associates developed an automatic system that mixed dialysis fluid concentrates with tap water in the proper proportion. Now each unit could be more compact and efficient; a single dialysis machine serviced five beds. The team also eliminated the need for a separate blood pump (the pressure of the patient's own bloodstream was enough to power the procedure), and simplified the machine's operations to the point where patients could routinely run the dialysis themselves.

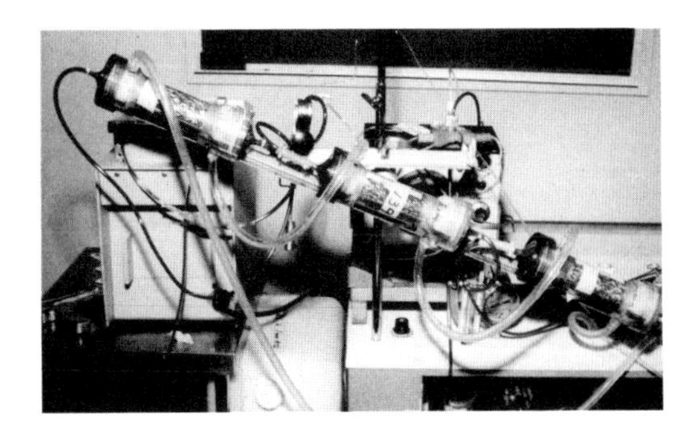

The treatment's limited availability, however, raised dramatic ethical questions. Since the demand was far greater than the number of beds available, and since the treatment was expensive, about $10,000 annually, an anonymous group of community members was empaneled to determine which applicants would be eligible to receive treatment. The panel's power over life and death was the subject of considerable publicity, including an hour-long 1965 NBC documentary narrated by Edwin Newman ("Who Shall Live?") and a lengthy *Life* magazine article by Shana Alexander (November 9, 1962). Scribner, Babb, and Quinton developed the first portable units, which made home dialysis possible, because of this limited availability. Home treatment freed patients from the need to spend lengthy, expensive periods in the hospital several times weekly. The first patient on the portable machine was a Seattle high school student, Caroline Helm, who had been rejected by the committee for receiving in-hospital dialysis but who began a successful course of home dialysis in 1965.

In keeping with a common practice in medical technology, no patents were sought on the dialysis equipment. The Scribner Cannula not only led to Seattle's pioneering center for the treatment of kidney disease, but also helped lay the foundation for several other related medical inventions: the Hickman Catheter for chemotherapy, created by Dr. Robert Hickman of Children's Orthopedic Hospital; the peritoneal dialysis machine devised by Dr. Henry Tenckhoff, now of Group Health Cooperative of Puget Sound; and the "artificial gut," a cannula for feeding nutrients to patients with terminal bowel disease, developed by a team headed by Scribner.

Seattle Foot

1985

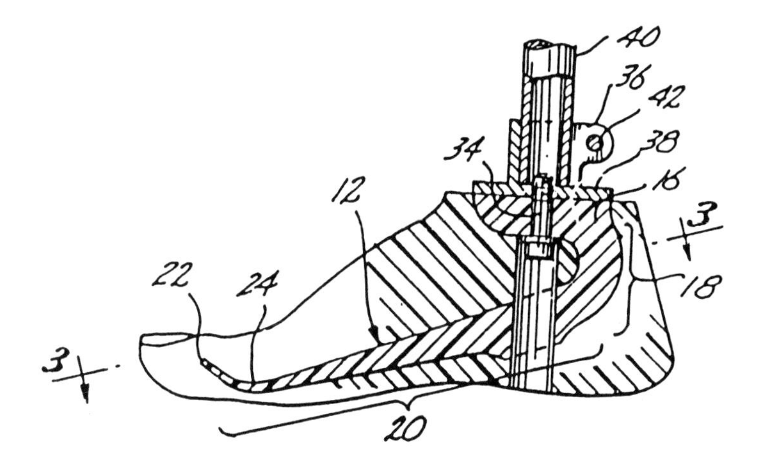

Move over, Long John Silver: here's the Seattle Foot, tough enough for any situation.

In 1988, the Du Pont company produced a striking television ad showing athlete Bill Demby, who lost both legs below the knee in Vietnam, playing a fast-paced game of pickup basketball. Only at the end of the commercial, when Demby sits down and unstraps his two artificial feet, is it apparent that he has been bounding around the court on false limbs.

Demby was one of the first amputees to use the Seattle Foot, a revolutionary prosthesis introduced in 1985 that has quite literally put a spring in the step of tens of thousands of lower-limb amputees. (Du Pont chemicals are used in its manufacture.) The Seattle Foot was the first element to be developed in the Seattle Prosthetic System, which includes an artificial ankle and shin as well as the first practical use of computer-aided design for custom-fitting prostheses to amputees. An artificial knee is currently in the design stage.

Created by a team of engineers, doctors, and designers at the nonprofit Prosthetics Research Study Lab under the auspices of the Veterans Administration and the University of Washington Department of Kinesiology, the Seattle Foot received a Presidential Design Achievement Award in 1984 and was a recipient of Washington's Governor's Award for New Products in 1990. Over 70,000 are now in use in the United States alone, and an outreach clinic established in 1989 in Hanoi is producing over 100 a month for the tens of thousands of people in Vietnam who were wounded during the war.

Among the distinguishing characteristics of the Seattle Foot is that it can be worn while performing extremely vigorous activities — such as skiing, rock climbing, running, or lifting heavy weights — proscribed for wearers of traditional prostheses. This is possible, in large part, because its design incorporates a patented spring, called a "monolithic keel," made of a strong, light, malleable, vibration-dampening material called Delrin. The keel stores the energy created when the foot steps down (roughly equivalent to body weight when walking but exceeding body weight by two to three times when running). The energy is then released at the

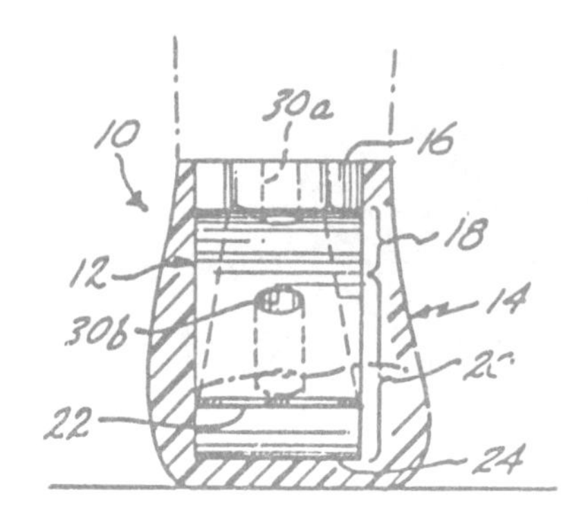

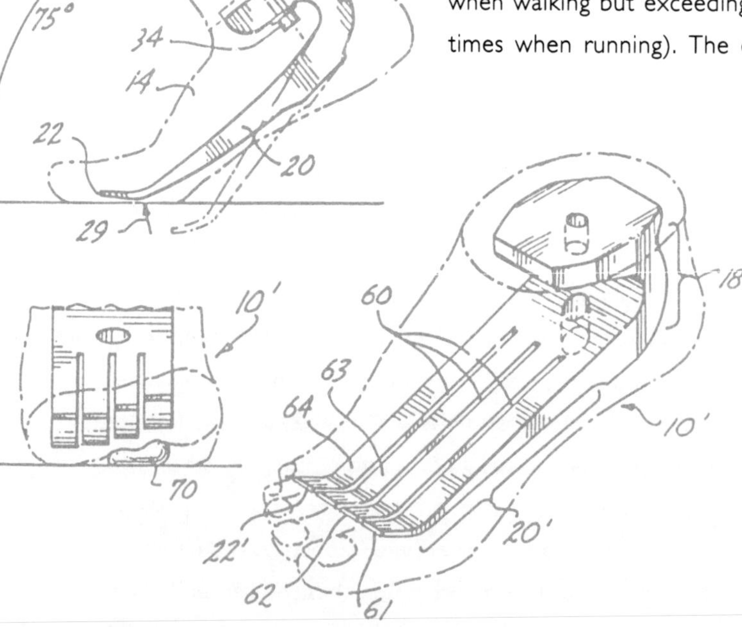

"ball" of the foot as the heel leaves the ground, giving a remarkably natural spring to the step.

The keel looks something like a thin, flat tongue. It sticks straight down from the ankle, curves back near the ankle, then moves forward over the arch and tapers out near the toes. Surrounding it is a protective pad made of foam, modeled for either men and women in a variety of weights and designed to look much more lifelike than previously available "blank" feet. The tip of the keel, near the toe, is reinforced at the bottom to prevent it from pushing out of the foot during especially rigorous activity. A bolt-hole runs through the foot vertically at the ankle, providing a means to attach it to an artificial shin. The foot is light (just over one pound), relatively inexpensive (about $175 for the basic foot), easy to replace (since the computer that custom-designs the foot for a given patient stores all revelant data for instant retrieval), and easy to manufacture. Computer technology has cut manufacturing time from weeks to hours. Senator Bob Kerry of Nebraska, a Vietnam-era amputee, was fitted with a foot in two hours on a Sunday afternoon while in Seattle for a political fund-raiser.

Many people helped create the foot. The project's guiding light was Dr. Ernest Burgess, a Seattle orthopedic surgeon who founded Prosthetics Research Study in 1962 and who continues to oversee its operation. The original keel was designed by an Auburn-based engineer, De Vere Lindh. Designer David Moeller and Don Poggi, an engineer and the CEO of Model and Instrument Development, the Seattle company that manufactures the foot, made it more flexible and easier to manufacture. Jon Harlan and David Boone wrote the software that lets prosthetists create custom limbs for their own patients. The Seattle Lightfoot, a modification designed by David Firth, further reduces cost and

weight, and because its keel is smaller, makes the foot available to Symes-level amputees, who retain their ankle bone and heel.

Several offshoots have emerged from the technology used to create the Seattle Foot. A "mud pod" or "altering foot," which looks something like a high-tech black plastic peg leg, was developed for use by Vietnamese workers in rice fields, where a normal artificial foot could become easily stuck in mud. Testing shows it is also good for fishing or for use in other rough terrain. De Vere Lindh, the designer of the original prosthesis, has used a similar spring technology to create a plastic horseshoe, the Seattle Shoe, especially for racehorses, jumpers, hunters, and quarter horses. Designed to prevent injuries and help injured horses return to the track more quickly, the Seattle Shoe absorbs up to 98% of the impact energy that would otherwise be passed through the horse's leg.

One offshoot of the Seattle Foot's technology is a vibration-dampening shoe that prevents injury to racehorses.

The Empathy Belly

1985

How would you feel, guys, if *you* were the ones who got pregnant?

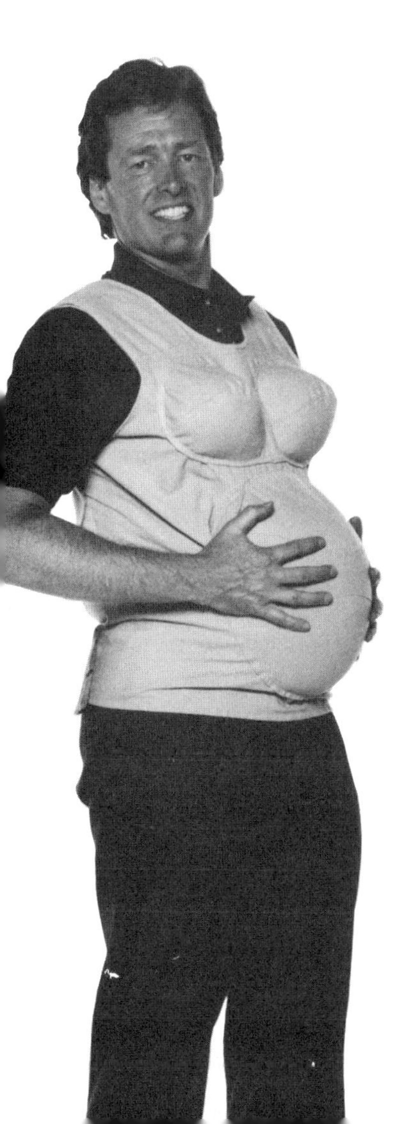

You've heard the jokes. If men got pregnant, they'd find a cure for stretch marks. Morning sickness would be America's number-one health problem. Men would stay in bed the entire nine months. They'd be *eager* to talk about commitment. And all methods of birth control would be 100% effective.

The Empathy Belly, invented by Redmond social worker Linda Ware, is a corsetlike garment that mimics 20 of the physical symptoms typically encountered by a woman in the last six months of pregnancy, such as extra body weight, enlarged breasts, and acute pressure on the lower back and bladder. The 33-pound Belly is made of canvas and features a tight rib-belt that constricts the lungs, a vinyl bladder filled with 11 pounds of warm water to simulate a distended abdomen, one two-pound weight suspended within the water to simulate fetal movement, a bladder pouch (six pounds of buckshot) that accurately re-creates the position of the fetal head pressing against the "mother's" real bladder, and two seven-pound lead balls that push against "her" ribs and internal organs.

Wearing the Empathy Belly doesn't provide an ersatz mom with some of the most uncomfortable symptoms of pregnancy, such as hemorrhoids, indigestion, constipation, shooting pains, or hormonal changes. Still, within half an hour of using the device, expectant "users" invariably report considerable physical difficulty: heightened blood pressure and pulse, altered posture, low backache and pelvic tilt, the need for frequent urination, a wobbly sense of balance, shortness of breath, fatigue, and a generally hot and sweaty demeanor.

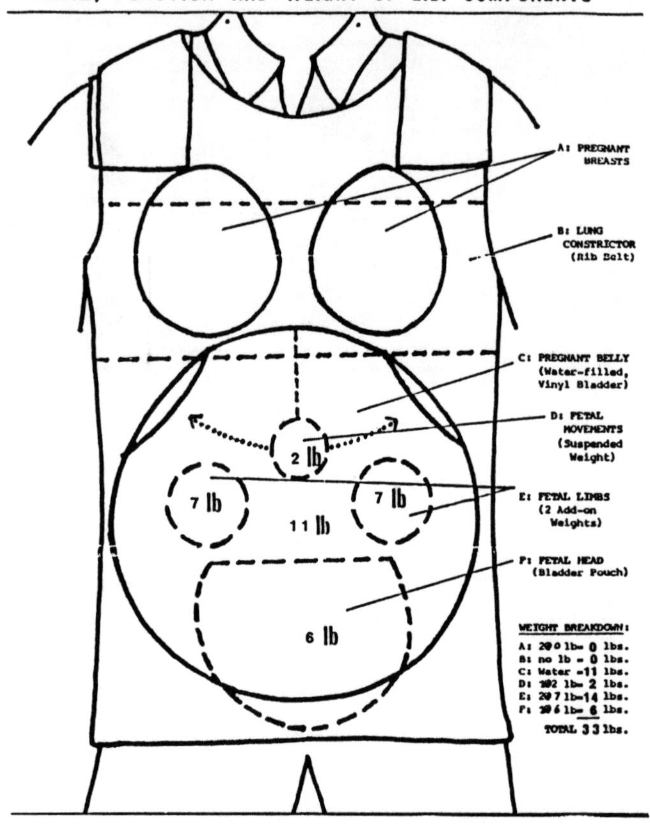

The Empathy Belly accentuates "couvade syndrome," a genuine condition of sympathetic pregnancy often felt by future fathers. Couvade syndrome — characterized by symptoms such as swollen ankles, weight gain, backaches, and upset stomach — is said to be felt by one out of five dads-to-be.

The point of all this pain and discomfort, Ware says, is twofold: to convey the negative consequences of teenage pregnancy and to increase the sensitivity of expectant fathers toward their wives. Ware's main focus is on sex-education classes for teenagers, some one million of whom, about one girl in 10, become pregnant every year in the United States — the highest rate in the industrialized world. The Empathy Belly shatters romantic illusions about teen pregnancy, and shows that "it takes maturity, effort, and support to expe-

The heavy tummy, tight rib cage, and pendulous breasts are fake, but the discomfort is real.

rience parenthood in a positive way. It's not just about having this cute, cuddly little thing to play with." At the very least, Ware hopes the Belly's use in high-school sex-education classes will get teens talking about the seriousness of birth control. One 16-year-old Virginia high school student (male), quoted in a *Washington Post* article, said: "At first, all I was really thinking about was sex. What a big change. [The Belly] ties you up. It'd be hard just getting in and out of a car. If you think about sex with a girl, you have to think about making a girl pregnant." The Belly is also used in prenatal classes at hospitals and other institutions to give prospective dads greater understanding of what their spouses are experiencing, to "demonstrate the need for specific shows of support" as well as the need to give practical, day-to-day help to pregnant women.

The idea for the Empathy Belly came out of Ware's experience as a childbirth counselor: "For fifteen years, my clients have been pregnant women and expectant fathers, and for fifteen years I've been finding that, despite all the discussion, men still weren't always *getting it* about pregnancy." While living in Eugene in the early 1980s, Ware saw a striking poster in a March of Dimes office that showed a "pregnant" teenage boy with the slogan, "Would you be more careful if it were you who got pregnant?" Ware was familiar with disabilities simulation — the use of props such as blindfolds or wheelchairs to gain understanding for the needs of the disabled — as well as with such learning environments as computerized flight simulation and the programs developed by highway-safety departments to simulate automobile crashes.

She began thinking about ways to simulate pregnancy and working on a prototype Belly in 1982. While living in Anchorage, Alaska, Ware field-tested it with students in her pre-natal classes. ("Just being pregnant myself wasn't enough. I consulted thousands of women about what they felt and experienced during pregnancy, and it was seriously field-tested by about 50 women.") Ware patented the Belly in 1985 and moved to the Seattle suburb of Redmond specifically for proximity to manufacturers, attorneys, and other resources necessary to market her product. It took four years to develop a marketing strategy and clear the hurdles set by the Food and Drug Administration, which regarded the device as a medical product, and the National Institute for Occupational Safety and Health, which set maximum safe lifting loads and standards to prevent such potential problems as back injuries. The first Belly was sold in 1989.

The Empathy Belly gets a lot of attention in the popular media. It has been featured in *People* magazine, on the *Donahue* and *Geraldo* shows, and in hundreds of other radio and TV shows, newspapers, and magazines. It has also been the occasional target of journalistic fun, such as Judith Stone's remark in the November 1990 issue of *Discover*: "Surely you gestate!" But the Empathy Belly was designed and is being marketed as a serious teaching tool. It is sold only to bona fide groups such as city and state health departments, junior- and senior-high school teen-pregnancy programs, Planned Parenthood chapters, and hospitals and medical schools. Accordingly, it is priced for institutional, not personal, use: a complete package retails for $645, including the Belly, training manuals, a maternity smock, videos, charts of uterine growth, and informed consent releases. Despite the high price tag, over 500 packages were sold in the first 18 months by Birthways Childbirth Resources, Inc., the company Ware formed to market the Empathy Belly worldwide.

Food and Other Poultry Matters

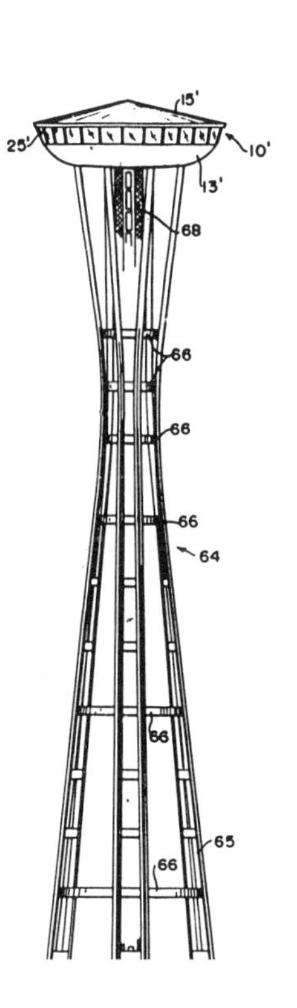

Evaporated Milk

1899

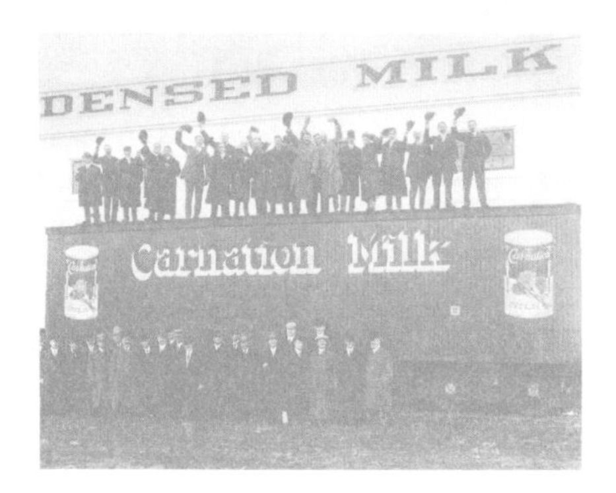

In the autumn of 1899, a small company in Washington's Kent Valley began supplying the world with something new: unsweetened, sterilized milk in cans. Carnation Sterilized Cream, as it was then known, soon became a household staple, especially among American housewives who liked the convenience of using it as a baby food. By the end of World War I, evaporated milk had achieved international status, helping to nourish a devastated generation of European children. It was as if, in the words of a welfare worker in France, "every child in this region is a foster-child of American canned milk."

Carnation was the result of a collaboration between two unlikely partners: a Midwest grocer and a cantankerous Swiss dairyman-inventor. E. A. Stuart, the "Stuart" of Seattle's landmark White-Henry-Stuart Building, was a Quaker, born in North Carolina and raised in rural Indiana. As a teen, Stuart had been so badly crippled with rheumatism that he was forced to use crutches, but eventually he regained his health through a variety of farm, railroad, and clerking jobs

E. A. Stuart was tireless in his efforts to promote the evaporated milk invented by his colleague John Meyenberg. Meyenberg used high heat to sterilize the milk, instead of adding sugar as a preservative (as in the condensed-milk formula patented by Gail Borden in 1856).

in the Midwest. By his early 20s, Stuart was the owner of a successful grocery store in El Paso, Texas.

Swiss-born John Meyenberg was a dairyman who had developed a product superior to existing brands of condensed milk. Unable to interest European investors in his new method, Meyenberg came to America in 1884 and was awarded U.S. patents that same year for both his process and his machinery. He settled in the dairy country around St. Louis and built the world's first evaporated milk plant. Under

One of Carnation's big selling points was its up-to-date and thoroughly sterile canning process.

new management, it eventually became the Pet Company, but Meyenberg was not able to make it financially successful.

In time, a few cans from Meyenberg's first factory, the Helvetia Milk Condensing Company, found their way to the shelves of E. A. Stuart's store in El Paso. Stuart and his wife tested the new product on their infant son and loved it. Stuart was so enthusiastic, in fact, that he contacted Meyenberg, who by then had relocated to California, and proposed a joint production venture with a third partner named Tom Yerxa. The Pacific Coast Condensed Milk Company (later Carnation Milk Products) was formed in 1899 when they bought out a small, failed condensed milk company near Kent.

Business was rough at first. Stuart and Yerxa knew nothing of the dairy business, old Meyenberg was intensely jealous of his secrets, no one knew a thing about the Northwest,

Elmer O. Pearson, a Swedish emigrant, settled in Seattle in 1907 and became a chemical engineer specializing in adhesives. In the 1930s, he was working for Borden Milk when he developed the formula for — you guessed it — Elmer's Glue. Pearson signed the rights away to Borden and never made any money from his invention.

the Kent plant was in serious disrepair, and Meyenberg's process was far from foolproof. By the end of the first year, the company was $140,000 in debt. Yerxa departed soon after, but Stuart and Meyenberg persevered. Their fledgling firm packaged its Carnation Sterilized Cream in distinctive red-and-white cans, 16 ounces for a dime. (Later it became known as "evaporated milk" when new legislation stipulated that "cream" had to contain 18% butterfat, a condition not possible with a canned product.) The brand name was chosen by Stuart, a flower fancier, who had seen the word "Carnation" on a box of cigars in a tobacco-shop window. Stuart was tireless in his efforts to promote Carnation, conducting personal demonstrations for shopkeepers and housewives, sending it to the Yukon (where it became a staple with prospectors and miners), even shipping it to Japan and back to prove it could withstand months of storage.

One day in 1907, Stuart was describing the bucolic Snoqualmie Valley, where his experimental farm was located, to Helen Mar, a young copywriter at the Chicago advertising agency he used. In a flash of inspiration, Mar came up with a phrase destined to become part of America's collective consciousness: "Carnation Milk...from contented cows."

Sang this 1915 booklet, "The most particular housewife would be delighted with the sweetness of every milk can and every utensil used in connection with the preparing of Carnation Milk."

Chicken Paraphernalia

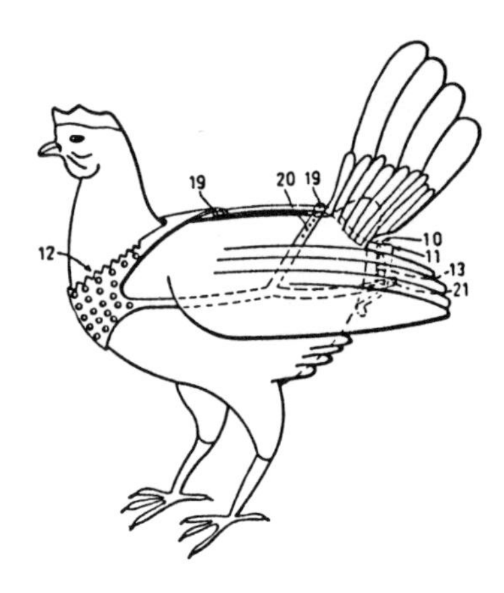

Several remarkable chicken-related inventions are credited to Northwesterners. One is the "mechanical hen" invented in 1907 by Gustaf Almstrom and Peter Christenson of Northport, Washington — a gadget for dispensing hard-boiled eggs in restaurants. When the diner pressed a lever, the metal hen's head would lift and the beak would open; the creature then cackled and a ready-to-eat egg dropped from the appropriate orifice.

In the early 1920s, a Centralia, Washington, man named Frank Swayne invented a trap nest for chickens. This device allowed farmers to isolate laying hens so that the number of eggs they produced could be readily noted. The results then told farmers whether to leave a particular chicken alone or to make it that evening's main dish. The invention enjoyed moderate financial success.

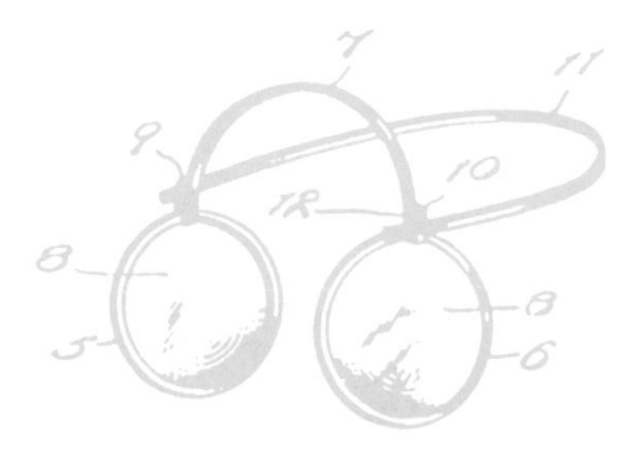

An anonymous Seattle inventor is credited with inventing protective muzzles for chickens. These handy items kept aggressive fowl from pecking other hens, but swung safely out of the way when the head was bent for eating.

In West Seattle during World War I, Lillian van Ornum Farmer, the mother of actress Frances Farmer, crossed a Rhode Island Red, a White Leghorn, and an Andalusian Blue to create a "red, white, and blue" chicken, which Farmer termed the *Bird Americana*. When the patriotic fowl made headlines across the country, Mrs. Farmer lobbied seriously but unsuccessfully to have it made the national bird.

Almstrom and Christenson's mechanical egg-dispensing hen (top) was in the same fowl spirit as the eyeglasses for chickens patented in Tennessee in 1903 (above and right).

Innovative Junk Food

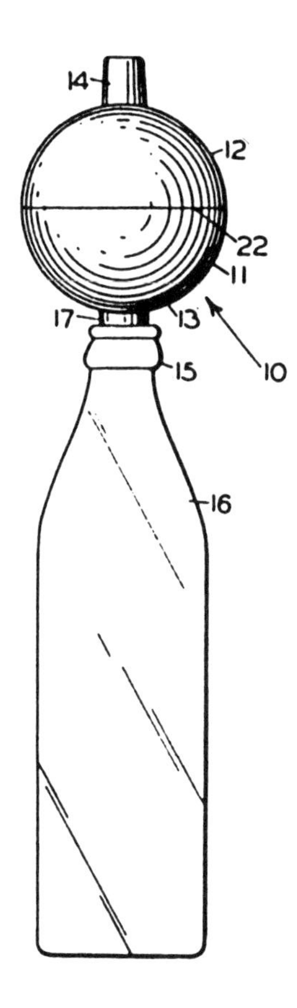

This "Container Attachment for Pop Bottles," marketed as the Fizz-Nik, was filled with ice cream to make an instant ice-cream float.

Northwest inventors have not been lax in their contributions to the ever-changing, brave new world of junk food. One early arrival was a machine for rolling ice cream cones, patented by F. Bruckman of Portland in 1902; up till then, the task had been done by hand. In 1932, another Portlander, Si Berry, began selling his popular Siberrian Cream Frozen Treat, an ancestor of today's soft ice cream, at his many "100 to One" shops. (One of Berry's machines is on display at the Oregon Historical Society Museum.) In the early 1960s, Seattle's kids were abuzz about the Fizz-Nik, an "instant ice cream soda" device invented by Glenn Chambers of Longview and distributed by Hays Merchandise of Seattle. It consisted of a pull-apart plastic sphere with two straws sticking out. Mom put a scoop of ice cream inside, closed the sphere, inserted one straw in a glass of pop and put the other in Junior's mouth. "Portable Ice Cream Soda Fad Set to Sweep Nation," the Seattle *Argus* proclaimed in 1961, but its impact proved to be only regional.

In the mid-1970s, Carl Gerdlund of Warren, Oregon, patented a device for automatically installing handle sticks in corndogs. In 1976, a flamboyant Seattle bus driver named Bob Allen invented a Rube Goldberg device that would split an Oreo cookie neatly in half, so that the inside cream could be licked off without damage to the cookie halves. Allen's device sent the cookie rolling from a slot atop a "splitter box" down a series of platforms to rest at the door of a conveyor, which then released the cookie to a guillotine-like apparatus that ensured a perfect split. His invention came complete with a transistor radio to keep the prospective eater entertained during the machine's lengthy splitting process.

While on the subject of nonnutritive food, mention should be made of a few of the classic Northwest candies. Though not inventions *per se*, they are remarkably durable regional creations. Almond Roca, the crunchy, chocolate-covered toffee candy, was made famous by its Tacoma manufacturer's slogan, "Brown and Haley makes 'em daily." The recipe for Frederick & Nelson's Frango Mints is credited to two Seattle candymakers of the early 1920s, Ray Clarence Alden and Joseph Vinikow, the latter a Russian who operated a confectionary called the Parisian Candy Company. Aplets and Cotlets, longtime favorites at roadside stands in Eastern Washington, were created in 1920 by brothers Marcar Balaban and Armen Tertsagian, who owned Liberty Orchards in Cashmere. They based the chewy treats on an Armenian jellied candy called *rahat locum* or Turkish delight, traditionally made with rose water and essence of orange but modified by the Liberty Orchards brothers to include apple or apricot juices.

A final candy note: At one time, Tacoma was quite a center for chocolate manufacture, and Forrest Mars of Mars Bars fame got his start in candymaking there. Unfortunately, Mars went bankrupt in the City of Destiny; it wasn't until he moved to the East Coast that he struck it rich.

DOG TOOTHBRUSH

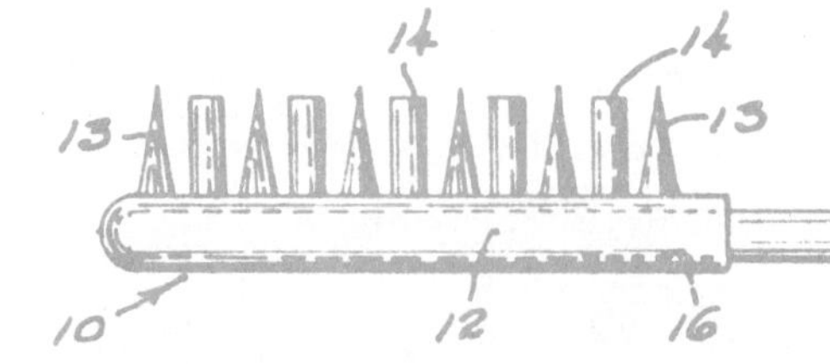

In 1961, a Seattle real estate agent and dog fancier named Bird A. Eyer invented a dog toothbrush. The brush was made of alternating rubber points and rubber cylinders; the latter held and applied toothpaste as the brush massaged the gums and teeth.

Revolving Restaurant

1959–1961

In 1959, Eddie Carlson, president of Western Hotels and one of the primary boosters of the upcoming Seattle World's Fair, visited the "TV Tower" restaurant, built atop a television tower in Stuttgart, Germany. He returned to Seattle with the idea that a similar tower could be built as a lasting and provocative symbol for the fair's theme, "Century 21," and presented the idea to other Seattleites involved in the exposition's planning. Jim Douglas, Century 21 vice-president and president of Northgate shopping mall, recommended that they invite Seattle architect John Graham Jr. to participate.

Graham was then in the middle of building the Ala Moana shopping mall in Honolulu. The centerpiece of this project was an otherwise unremarkable 20-story office building that featured an idea Graham had invented: a revolving restaurant that turned slowly on its axis to give diners a panoramic view. Graham suggested that a similar idea might work for the new Seattle tower. The Century 21 executives were enthusiastic, and Graham's team, which included architects Nathan Wilkinson and Art Edwards and designer John Ridley, began preliminary studies. No one is quite sure who coined the tower's name, but it was probably someone on Graham's staff. In any event, the tower and its restaurant were to gain lasting fame — and a permanent place in the hearts of Northwesterners — as the Space Needle.

Many ideas were proposed for the tower's shape, including a far-out "helium balloon" design and a stark 500-foot concrete tower that one observer said resembled "the Washington Monument with a wheel on top." The design that

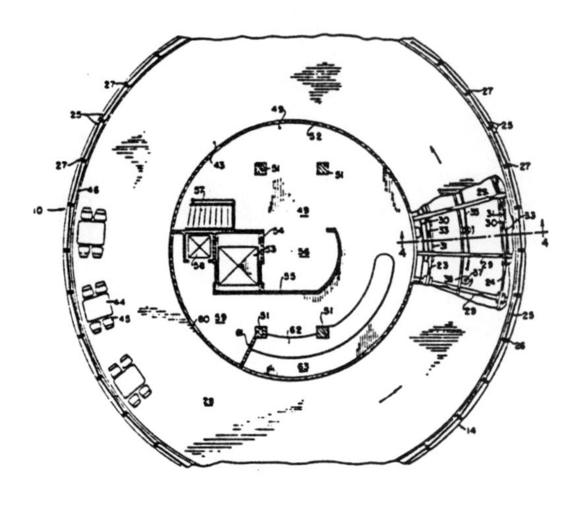

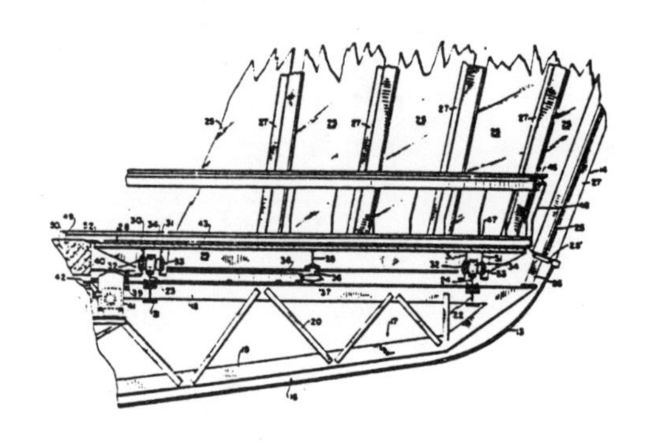

The unique characteristics of the Space Needle have led to a number of unusual touches, such as color-coding each quarter of the circle to ensure that waiters and waitresses don't misplace their customers.

was finally chosen was more graceful and more structurally sound: a tripod form with a slim central column to support three elevators. It was developed after Graham brought in some last-minute help: Al Miller, a University of Washington professor of engineering, and Victor Steinbrueck, a University of Washington professor of architecture (and later the spearhead of the movement that saved the Pike Place Public Market from destruction by urban renewal).

Graham's patent application for a "restaurant with rotating floor," filed in 1961 and granted in 1964, featured drawings showing the restaurant both atop an Ala Moana–style office building and on a tripod similar to the Space Needle's ultimate form. The application describes it as "a restaurant of novel construction, which is to be erected at a considerable elevation on a supporting structure on the top of a building, or on a tower built for the purpose. The new restaurant has an outer wall formed mainly of transparent panels and, in order that the patrons may enjoy the panoramic view afforded by the elevation, the dining area has a rotating annular floor equipped with tables and chairs and having its outer edge lying close to the outer wall. The central area of the annular floor is closed by a stationary floor, which

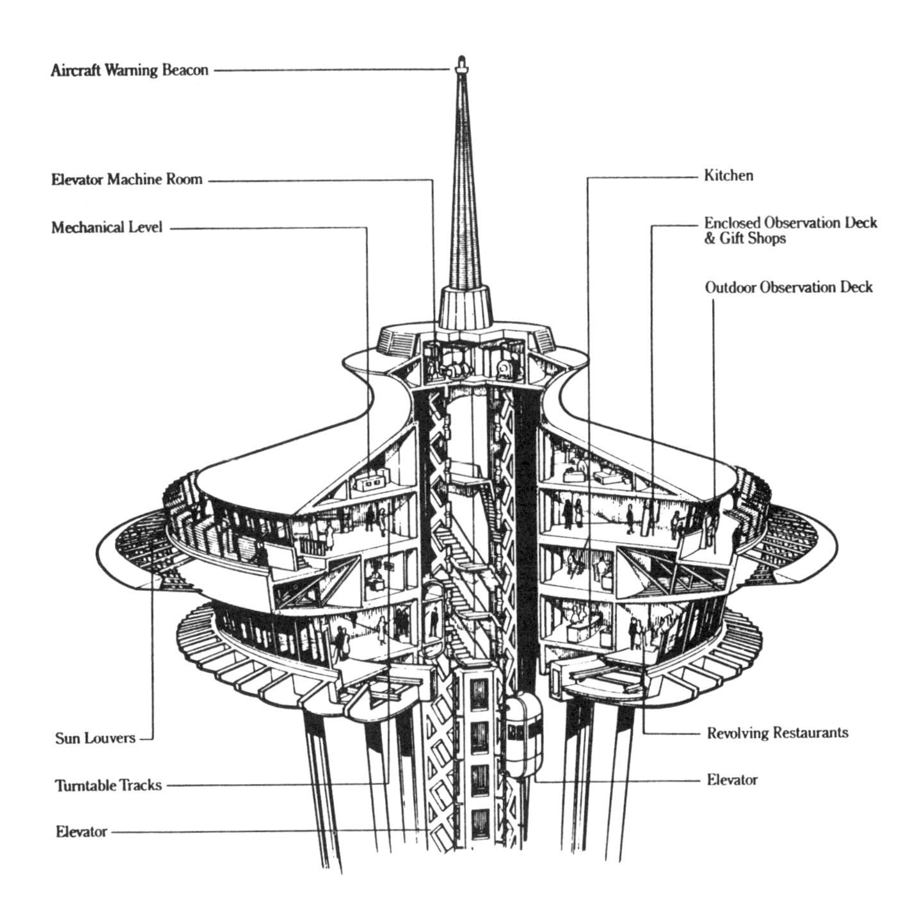

has one or more openings for the means of access to the dining area."

The final design put the restaurant at an elevation of 628 feet, with an observation deck and a television tower that raised the structure's total height to 730 feet. The restaurant had kitchen facilities on the same floor and space for 280 diners. The structure's center of gravity was placed near ground level, and extensive testing in a UW wind tunnel showed it would withstand storms of up to 100 miles per hour, well above the highest recorded wind speed in Seattle. (Two hundred and fifty tons of reinforcing steel, 72 thirty-foot bolts, and 5,600 tons of concrete went into its foundation alone.) The construction of the tower itself was

a remarkable feat of engineering on the part of consulting engineer John Minasian, contractor Howard Wright, and the hundreds of steelworkers and other workers involved. The foundation was begun in April of 1961 and the basic structure completed by December of that year, without a single fatality and at a final cost of nearly $4 million.

The restaurant's floor makes a 360-degree journey every hour — 5,840 times per year. It uses a simple ring gear with a small cog, geared low so that a single one-horsepower motor is sufficient to drive it. (The Ala Moana "prototype" used two large motors, which had synchronization problems that gave the restaurant a jerking motion.)

Graham played a major role in another architectural innovation, one that has had permanent and far-reaching implications: he designed and codeveloped Northgate, the world's first regional shopping mall. Northgate's basic plan — a single, narrow aisle with small shops on either side, large "anchor" stores on either end, underground service access, ample and accessible parking — is the paradigm for similar malls around the world.

MARASCHINO CHERRIES

Maraschino cherries originally obtained their sweetness and potency from being packed in a brine made of fermented cherry juice and herbs. When Prohibition halted the import of European maraschino liqueur to the United States, Oregon cherry growers (who supply nearly two-thirds of the nation's maraschino cherries) were stuck. Help came in 1925 from E. H. Wiegand, a food technologist at Oregon State College (later OSU). He developed a method of brining that reduced the packing time from two years to a few days, an inexpensive method that has since been adopted universally.

Triploid ("Sexless") Oysters

1985

Romantic sponges, so they say, do it
Oysters down in Oyster Bay do it
Let's do it
Let's fall in love.
— Cole Porter, "Let's Do It (Let's Fall in Love)"

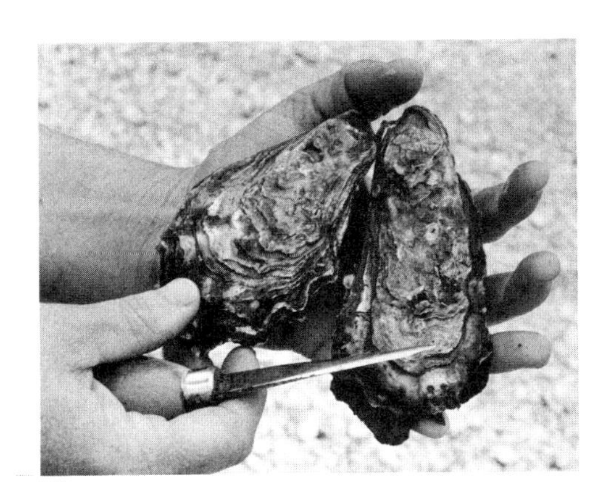

Triploid oysters are oysters that have been genetically engineered to be better-tasting, even in summer, when oysters are traditionally unpalatable. They account for roughly 10% of all the oysters produced in the Pacific Northwest, which is the second largest oyster-producing area in the country after the Gulf States. Developed in the mid-1980s by a team of researchers at the University of Washington School of Fisheries, the triploid oyster became the subject of a landmark test case when a Seattle law firm sought to make it the recipient of the first patent for a genetically mutated animal. The oysters' supporters lost the case, but their application led the U.S. Patent Office and the U.S. Court of

Appeals to declare that genetically altered higher animals can indeed be patented. This opened the floodgates for a surge of advanced research and patent applications by biotechnologists and genetic engineers. The far-reaching implications of the court's decision are only now beginning to be realized.

Triploid oysters differ from normal oysters in that they have three sets of chromosomes in each cell instead of two. (Most sexually reproducing organisms, oysters and humans included, are diploid, with two sets in each cell.) The formation of the gonads in a triploid is severely inhibited, so that carbohydrates that would normally be converted to lipid (eggs) or protein (sperm) are instead stored in the main body of the oyster. The result is a faster-growing, sweeter-tasting, better-textured product. (It is also more nutritious: higher in protein and lower in fat.) Media attention at the time of its introduction led to the triploid's characterization as a "sexless," "eunuch," or "sterile" creature. Technically, it is none of these things. It simply spends much less energy on sexual reproduction than does a normal oyster: roughly 10% to 25% of its total body weight, as opposed to the normal rate of 60% to 80%. "Sexually retarded" or "sexually inhibited" would be more accurate terms, though somewhat less colorful. Despite the triploid's sexual inhibition, its advocates are quick to point out that the oyster's mythical aphrodisiacal qualities are not, as far as they can tell, affected by genetic mutation. Triploidy (the presence of the triploid state) as a potential tool for commercial aquaculture was first proposed in the early 1970s, and some preliminary research was done at the University of Maine in the early 1980s on American oysters (*Crassotrea virginica*), which are native to that region. But the concept's potential usefulness was not realized by the relatively small scale of Northeast oyster hatcheries, perhaps

The old caution against eating oysters in months with "R" in their names is a myth. The belief that oysters are unhealthy in summer stems from the life cycle of the European genus *Ostrea*, which broods its young inside itself and is filled with tiny shells in summer (crunching down on these is unpleasant, but not necessarily unhealthy). All North American varieties, including triploid Pacific oysters, are safe to eat year-round.

because American oysters are not as sexually active as the Pacific oysters (*Crassotrea gigas*) grown in the Northwest. (American oysters devote only about 30% of their bodies to reproduction, as compared with the Pacific oyster's 60% to 80%.)

Triploidy found favor, however, among the well-established hatcheries of the Northwest, where Pacific oysters have been farmed in stabilized beds since they were imported from Japan in the 1890s. Hatchery methods have been perfected in the Northwest because the region's coastal waters are too cold for the Pacific oyster to spawn naturally. The young must be reared under controlled conditions.

In 1983, Sandra Downing and Standish Allen, two researchers at the University of Washington under the supervision of Professor Kenneth Chew, repeated and extended some of the experiments in triploidy begun by the University of Maine team, of which Allen had been a member. The following year, with backing from the Washington Sea Grant and the Pacific Coast Oyster Growers Association, they began to apply their research to large-scale production.

Inducing triploidy in animals requires intervention during meiosis, the process whereby cells reduce their two chromosome sets to one before sexual union. In meiosis, chromosomes in each male and female germ are duplicated so that each temporarily contains four chromosome elements. The male cell splits into two spermatocytes, each containing half the chromosomes of the parent cell; these divide again, and the resulting cells contain half the chromosomes of their parent. In the female, the egg is released while retaining its four sets. When the sperm penetrates the egg's outer membrane the female divides, but does not produce four identical cells; instead it extrudes a small piece of cytoplasm, called a polar body, containing half the chromosomes of the parent cell. A second polar body containing one set of chromosomes is released during the female's second division, leaving the egg with one set of chromosomes. When this set fuses with the set contributed by the sperm, the result is the beginning of a normal diploid organism.

In an oyster this sequence can be manipulated by any of several methods, including hydrostatic pressure, heat shock, and chemical intervention, so that the egg retains, rather than extrudes, the second polar body. The egg then contains two sets of chromosomes, the sperm contributes another, and the result is a triploid oyster. Allen and Downing first experimented with hydrostatic pressure, which had induced triploidy in fish by reproducing the pressures found undersea. This method proved unsuccessful for oysters, however, since oyster eggs — which are about 40 microns in size, some 1,000 times smaller than fish eggs — experienced much lower survival rates at high pressures than the fish eggs.

Allen and Downing then applied a chemical called Cytochalasin B (CB), which was originally used as a positive control but which produced higher survival rates and triploid percentages. Incorporating CB into oyster production involves treating the gametes with the chemical and rinsing them when the second polar body formation takes place. This time frame is determined by factors such as water temperature.

Once it was proven that triploidy could be reliably induced and that it had considerable potential for large-scale production, the researchers were approached by an oyster-growing concern, Coast Oyster, about the possibility of patenting the process. Originally, only the technique of using hydrostatic pressure to induce triploidy was to be the object of the patent, but the researchers decided to try and patent the entire animal. (The use of chemical inducement was unpatentable because it had been previously published.) The University of Washington's Office of Technology Transfer and the Seattle patent-law firm of Seed and Berry were interested, in large part because of its potential as a landmark test case.

In 1987, the U.S. Court of Appeals' Board of Patent Awards and Interferences ruled that the triploid oyster was not patentable as a separate, unique animal, for two reasons. Some of the preliminary research had been done on the East Coast using American oysters prior to the UW team's efforts; also, triploidy sometimes occurs naturally in hatcheries. (One explanation is that when there is a great deal of sperm in the water, two sperms will sometimes fertilize a single egg, although this is rare in the Pacific oyster.) The court did, however, award a patent to Allen, Downing, and Jonathan Chaiton, a biologist at Coast Oyster, for the hydrostatic-pressure technique; more significantly, it issued its decision that higher animal forms in general were patentable. The animal that was eventually awarded the first patent for a "transgenic nonhuman mammal" was a mouse, genetically altered to facilitate cancer research, which was developed by Dr. Philip Leder of Harvard Medical School, and Dr. Timothy Stewart, a senior scientist at Genentech in San Francisco.

VARIOUS EDWARDIAN INVENTIONS

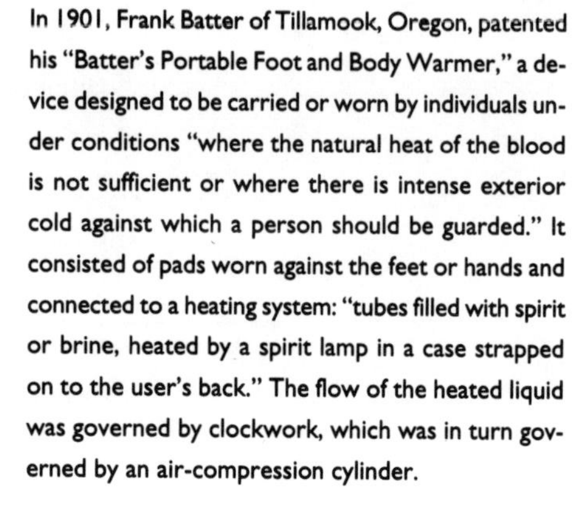

In 1901, Frank Batter of Tillamook, Oregon, patented his "Batter's Portable Foot and Body Warmer," a device designed to be carried or worn by individuals under conditions "where the natural heat of the blood is not sufficient or where there is intense exterior cold against which a person should be guarded." It consisted of pads worn against the feet or hands and connected to a heating system: "tubes filled with spirit or brine, heated by a spirit lamp in a case strapped on to the user's back." The flow of the heated liquid was governed by clockwork, which was in turn governed by an air-compression cylinder.

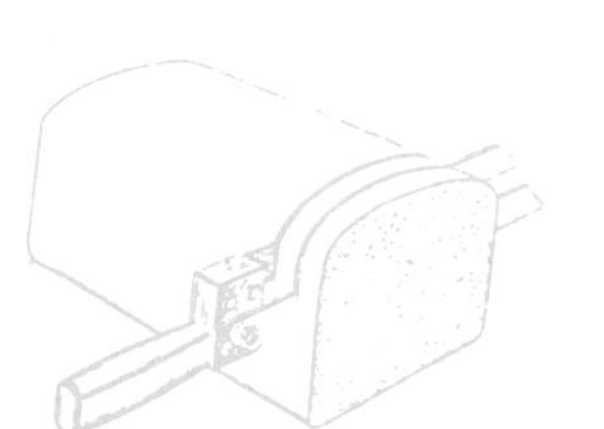

An "Improved Knife for Slicing Bread, Cake, and other Light Spongy Materials" was invented in 1904 by two brothers, John Calvin Bollinger and George Edward Bollinger of Olympia. It was a two-bladed knife with the blades arranged in step fashion on a single handle, so that more than one piece could be cut at a time. The difference in the height of the step was adjustable so that thick or thin slices could be cut, as desired.

The "Orthopaedic Exercising Apparatus" was designed to stretch the vertebral column and exercise and strengthen the muscles of the neck and upper body. It consisted of a cap bearing a pulley on the top, which was strapped to the patient's head. Handles were attached to a cord which was so arranged that by pulling down on them the patient could raise himself from the floor by the upward pressure exerted upon the head. It was invented in 1904 by John Kleinbach of Spokane.

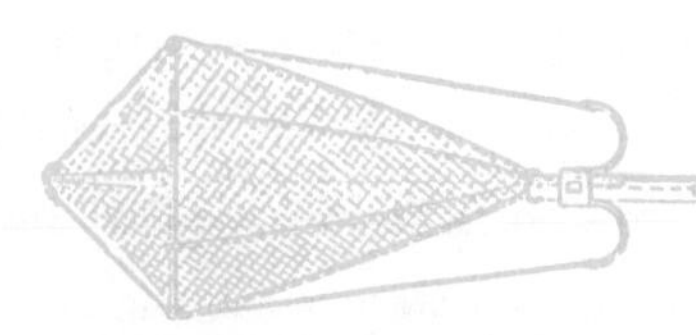

"Terletzky's Improved Insect Catching Device," invented in 1905 by Max Terletzky of Goble, Oregon, employed arrows fitted with open baskets or nets at their tips. These arrows were then fired at insects, and the baskets closed automatically when they reached the end of their flight, trapping their quarry.

Super Salmon

1940s–1960s

Perhaps no other animal symbolizes the Pacific Northwest better than the wild salmon. The salmon is beautiful, strong, and courageous. Nobody who has seen one fighting its way up a mountain stream, part of a journey that can cover hundreds or even thousands of miles, would disagree. It is also profoundly mysterious: one of biology's most enduring and tantalizing secrets is how salmon track their way backwards, years later, to find their exact spawning grounds. The salmon is endangered today, thanks in part to some of the inventions mentioned in this book, such as the fish wheel, the Iron Chink, and the purse-seiner Power Block, which allow us to take greater numbers of fish from the water. In the 1950s and 1960s, Dr. Lauren Donaldson of the University of Washington School of Fisheries developed strains of "super salmon" and "super trout" as one way of compensating for the trends that threaten to wipe out the wild species.

By careful selective breeding, Donaldson was able to produce rainbow and other types of trout, as well as coho, Chinook, sockeye, pink, and chum salmon, with much greater size, strength, rapidity of growth, and fecundity than wild species. (The process had never before been tried on salmon, because they were considered unyielding to biological or physiological manipulation.) Donaldson's trout and salmon had considerably higher rates of survival, carried out spawning cycles that were shortened by as much as half, and produced young at a rate up to 15 times greater than that of wild fish.

Donaldson grew up in a small town in Minnesota near

Dr. Lauren Donaldson's dream of selectively bred salmon that would be bigger, stronger, faster-growing, and more fecund than wild varieties is still a controversial issue between proponents of hatchery-raised fish and environmentalists concerned with the preservation of wild species.

the South Dakota border and graduated from the University of Montana in 1926, majoring in biology and chemistry. After four years as the science teacher, athletic coach, and eventually principal of Shelby (Montana) High School, he came to graduate school in the fisheries department of the University of Washington to pursue his interest in fish breeding. The resources at the school of fisheries were severely limited during the Depression, and Donaldson had to create hatchery facilities on his own: wooden troughs, piped-in running water at controlled temperatures, hand-mixed food. He was particularly interested in the proper nutrition of hatchery-raised anadromous fish: his 1939 doctoral thesis was entitled "Experimental Studies in the Nutrition of the Chinook Salmon with a Special Reference to Histological Changes in the Pancreas." (Anadromous fish run upriver to spawn; their opposite counterparts are catadromous. Histology is the study of microscopic changes in tissue formation.) Donaldson joined the school's faculty in 1941 and became a professor in 1948, a position he held until his retirement as professor emeritus in 1973.

In 1949, Donaldson began the project that became his major work: a completely artificial salmon run with the spawning ground right at the fisheries school, at the foot of the university campus, on Portage Bay between Lake Union and Lake Washington. The experiment would allow him to observe and control the spawning environment at close hand, but spawning had never been tried in an urban environment where the salmon would have to navigate, among other obstacles, the man-made Chittenden Locks that connect Lake Union with saltwater Elliott Bay and Puget Sound.

That year, Donaldson released 23,000 young Chinooks, with subsequent releases in 1950, 1951, and 1952. The Chinook's normal spawning cycle is three to five years, and in November 1953, 23 four-year-old fish — members of Donaldson's first 1949 group — came back up the concrete ladders at the foot of the university ponds and marked the beginning of his selective breeding process. By 1959, the percentages of fish homing to the university grounds not only exceeded normal expectations but had multiplied until they were 30 times the returns found in natural environments. Egg production in these survivors was high enough that the fry increased salmon plantings in the state tenfold.

While waiting for his annual returns, Donaldson experimented with other breeding factors. He accelerated the growth rates of fish by warming the incubation waters for eggs and fry, and he tried hundreds of combinations of special diets. The optimum diet proved to be 50% vacuum-dried fish meal, with the remainder kelp, a digest of whole salmon viscera, a complete vitamin mix, and shrimp waste. Rice hulls and wheat germ were also used on occasion. The effects of altered diet on Donaldson's fish were spectacular: the growth rate of test rainbow trout, for instance, increased fivefold by diet alone, and when diet was combined with selective breeding, increases in size were as high as 100 times those of wild fish.

Donaldson found that increased growth rates affected the length of spawning cycles. By 1966, his incubation and diet procedures had become so finely tuned that groups of sockeye eggs could be raised, released, and returned in three years instead of the normal five. Incubation and diet had caused the fish to grow much faster, and the two-year period normally spent by sockeyes in fresh water before migrating to the sea was virtually eliminated. After 12 years of selective breeding, rainbow trout needed only two years to reach maturity, a cut from four years. But the most striking difference in hatchery-bred rainbows was in their fertility: in 1973 a Donaldson rainbow, a female of the stock that had been cultivated for 40 years, produced 27,636 eggs — more than 20 times the number expected from fish living in natural streams.

Selective breeding, which can be so successful in developing an agricultural product such as Gaines wheat (see next page), is controversial when applied to fisheries research. Many fisheries biologists argue that genetic diversity is important for salmon because of their migratory nature. Agriculturists have a degree of control over the land in which grain is sown; it is felt that biologists, working within the context of a natural, constantly changing, and less controllable environment, must take greater care to maintain the variety and range of available genetic material. Critics argue that hatchery-bred salmon are shorter-lived and less tasty than their wild counterparts. They also point to considerable evidence that replacing wild salmon with hatchery-bred salmon runs can severely damage the environment. Nonetheless, government agencies remain generally in favor of hatcheries.

Gaines Wheat

1961

The variety of grain known as Gaines Wheat, developed by researchers at Washington State University in Pullman, helped revolutionize wheat production worldwide. By playing a pivotal role in the Green Revolution of the 1960s, which brought disease-resistant and highly productive strains of wheat, rice, and corn to millions of farmers in tropical nations, Gaines increased worldwide wheat yields by an average of 50%. Locally, it has doubled Washington State's wheat production and boosted Pacific Northwest wheat crops as a whole by over 25%. There are estimates that Gaines wheat adds about $100 million annually to the Northwest economy.

Wheat farmers in 1930s America faced a serious problem. They were accustomed to planting crops late in the spring to allow sufficient time beforehand for mechanical weed control. Newly introduced chemical weed killers made it possible to plant wheat earlier and thus reap bigger harvests. The problem was that the wheat now grew too tall, developed a heavier head, and fell over, or "lodged," in the heavy spring rainfall. Dr. Orville A. Vogel, a newly appointed Department of Agriculture scientist at Washington State College (now WSU), was given the task in 1931 of developing a variety of wheat that was not only disease-resistant but stiff enough to stand up to the spring rains.

From 1931 to 1961, Vogel undertook painstaking experimentation, crossing varieties of wheat that had promising characteristics, exposing them to dangers, and then carefully examining the results. The odds against success were staggeringly high. Vogel once estimated that scientists trying to

Dr. O. A. Vogel (left) inspects a planting of Omar wheat. Omar is a smut resistant variety bred by Dr. Vogel.

incorporate 10 genes into a new variety had one chance in 1,084,576 of producing the desired characteristics, and that even if successful the odds would be the same of realizing that such a plant had been produced. At various times, Vogel worked with 22 different genes in a single breeding effort; the odds of incorporating all 22 into a single plant were estimated at one in one sextillion (1:1,000,000,000,000,000,000,000).

In his early research, Vogel developed several varieties of wheat that worked well; they had stiffer straw and wouldn't lodge in spring rains. The farmers who tested them, however, had a tendency to pour on too much fertilizer. The plants grew too tall, became top-heavy, and fell over. The breakthrough finally came with a semidwarf variety called Gaines, which derived from a short-strawed wheat called Norin-10. Norin-10 was a Japanese breeding line isolated by Vogel's superior, Dr. S. C. Salmon of the U.S. Department of Agriculture, who was helping Japan get its agricultural stations restarted during the period following World War II. One night in March 1949, Vogel was playing poker with a colleague, Dr. Fred Elliott, when Elliott mentioned that he had a graduate student, Dick Nagamitsu, who needed some greenhouse experience. Vogel had just received a sample of the Norin-10 dwarfing-gene germ plasm, and Nagamitsu was put to work crossbreeding the new variety.

Twelve years later, in 1961, the Gaines hybrid, based on Vogel's original sample of Norin-10, was introduced to the public. Gaines grew on short, stiff straw, could be planted early in the fall, was relatively disease-resistant, and stayed upright even with excess chemical fertilizer or heavy rains. The variety was a huge success domestically, but its true impact came when Vogel sent samples to several major plant breeders, including Dr. Norman Borlaug, director of the Rockefeller Foundation's wheat-breeding program in Mexico.

Borlaug was developing a variety of Mexican dwarf spring wheat that could be planted in hot, low-lying areas, and he used Vogel's dwarfing gene to perfect his own work. Borlaug's efforts, which earned him the 1970 Nobel Peace Prize, helped launch the sweeping agricultural reforms known as the Green Revolution. The results of this program were spectacular: Pakistan, the second largest recipient of food aid from the United States, became nearly self-sufficient in cereals within a matter of years. India's production of wheat rose 50% between 1965 and 1969. The Philippines ended its policy of rice importation. Ceylon's rice crop increased 34% in two years. The list goes on and on; efforts toward agricultural self-sufficiency in virtually every South American, Asian, and African nation have been aided by Vogel's pioneering work.

President Gerald Ford presented a National Medal of Science Award to Vogel in 1976. The award stated: "Dr. Vogel has consistently recognized the value of subjecting his breeding material to a myriad of hazardous environments and disease complexes." Vogel, who dismissed fancy theories and liked to call himself "a dumb wheat farmer," put it slightly differently in a *Seattle Times* interview shortly before his death in 1991: "My approach was to beat the hell out of [the test strains] and take the survivors."

Automatic Leveler for Grain Combines

1942

R. A. Hanson's company, RAHCO, is a Spokane-based manufacturer of specialized construction equipment, a world leader in large-scale machinery for canal digging with over 100 patents. The firm was founded, however, on the strength of the invention that Raymond Hanson says is still his proudest achievement: a self-leveling device for combine harvesters. Combine harvesters are machines that both reap and thresh grain. One problem with them is the need to keep their working parts, such as the cutting knives and threshing machinery, on a steady horizontal level even when the crop being harvested has been planted on an incline.

Hanson was familiar with this situation, since he spent many of his summer hours towing a combine harvester behind a tractor, up and down hills on his father's wheat ranch in the Palouse country of southeastern Washington. He recalls, "Harvesters in those days did have a leveling device, but it was manually operated — just a mechanical device that jacked one leg up and down. The problem was that you couldn't do a very good job; you had to stand up and operate a lever while you were driving, and it didn't work out very well. It's very difficult, when you're on a hillside, to tell what's level; you tend to compromise with the hill, to always lean downhill some." During the summer of 1942, when he was 19 years old, Hanson was inspired to try to build a control that would automatically adjust the combine to compensate for inclines.

Hanson worked on the project over the winter of 1942–43 in his father's small shop, with a used college textbook

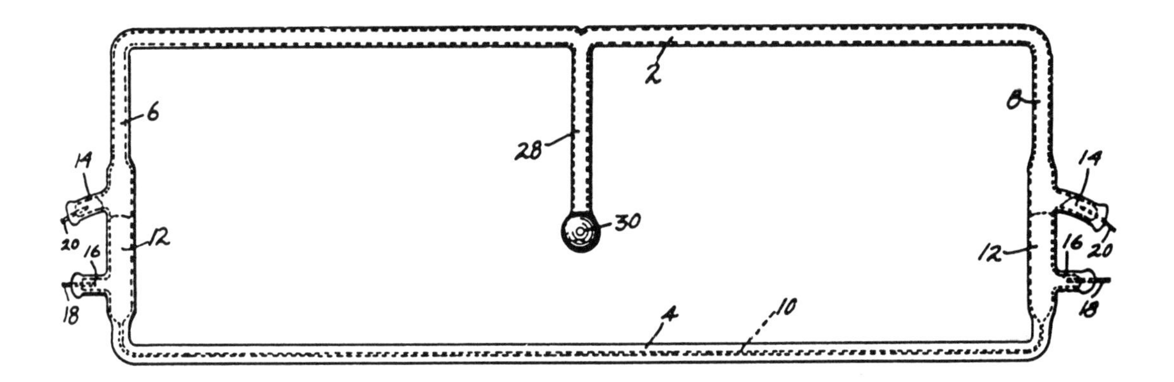

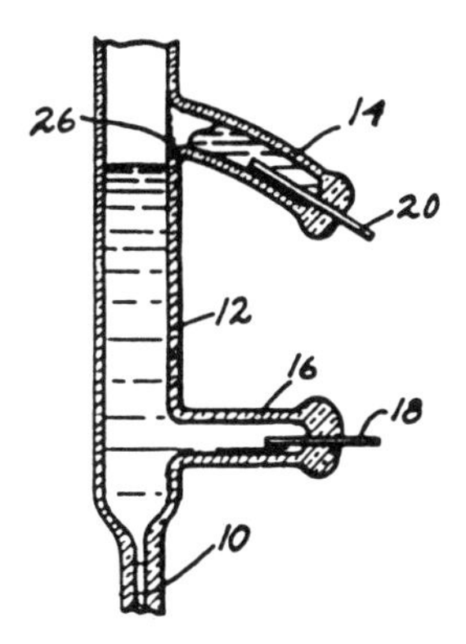

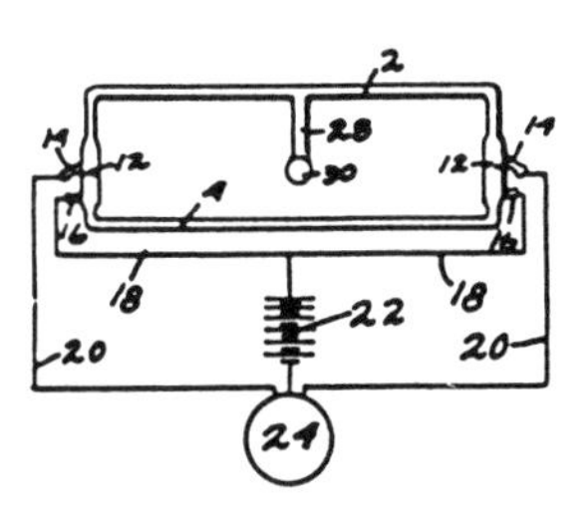

called *Electro-Magnetic Devices* — which he still has in his library — as his main reference. He rejected the use of a pendulum or a grooved track with ball bearings to control the degree of compensation for different slopes, both of which had been tried unsuccessfully by others in the past, and instead concentrated on building a frictionless mercury switch. He set the switch in the middle of an elongated, U-shaped metal tube, the legs of which were about 18 inches apart, with the connecting section forming a restricted passage for the mercury. He mounted his level on the underside of a harvester so that, as the mercury ran from one side to the other (with the restricted passage preventing it from rushing too quickly back and forth), it made or broke an electrical contact. This, in turn, activated the mechanism that adjusted the level.

Hanson, who was working toward a degree at the University of Idaho in mechanical engineering, made five of his self-levelers that winter and sold them to local farmers for $200 apiece. The next winter he hired a friend to help him, and in successive years they turned out 10 and then 20 levelers. Hanson's schooling was interrupted by Navy service during World War II, but by 1946, back in the Palouse, he had patented the invention, abandoned school to form the R. A. Hanson Company, and was making 200 levelers a year at $400 apiece. Although he only completed a year and a half at U. of I., he was awarded an honorary doctorate from the school in 1985.

Hanson says he was "predestined to be an engineer and inventor." His full name is Raymond Alvah Hanson; his middle name, a variation on Thomas Alva Edison's, was his mother's homage to the inventor. Hanson's father was a farmer and handyman who was, his son says, "the real inventive genius in the family. They called him Haywire because he could fix anything, even if he had to do it with haywire. He did all kinds of amazing things with practically nothing, hardly even a blacksmith shop to work with." Among the elder Hanson's inventions was a water-driven wood-cutting saw built when he was a 14-year-old farmboy in Peck, Idaho. Hanson recalls, "Back then they used to cut logs into four-foot lengths by hand, split them by hand, and then cut them twice more into sixteen-inch lengths on a buzz saw, usually powered by a Model T engine. My dad figured out a way to make a wood-cutting machine powered by a big water wheel, with about a thirty-forty-foot span; the log would come off rollers onto a bench, a saw would cut a sixteen-inch length off on the first pass, and it would advance another sixteen inches. It was much easier, and all automatic." But Hanson Senior never patented that machine or any of his other inventions. His son says, "He never made a dollar, never commercialized a thing. He just solved the mechanical problems of the neighbors for years and never took a dime. But I had a little Scotchman in me or something; I wasn't doing all that freebie stuff. My dad was the genius; I was just a better businessman."

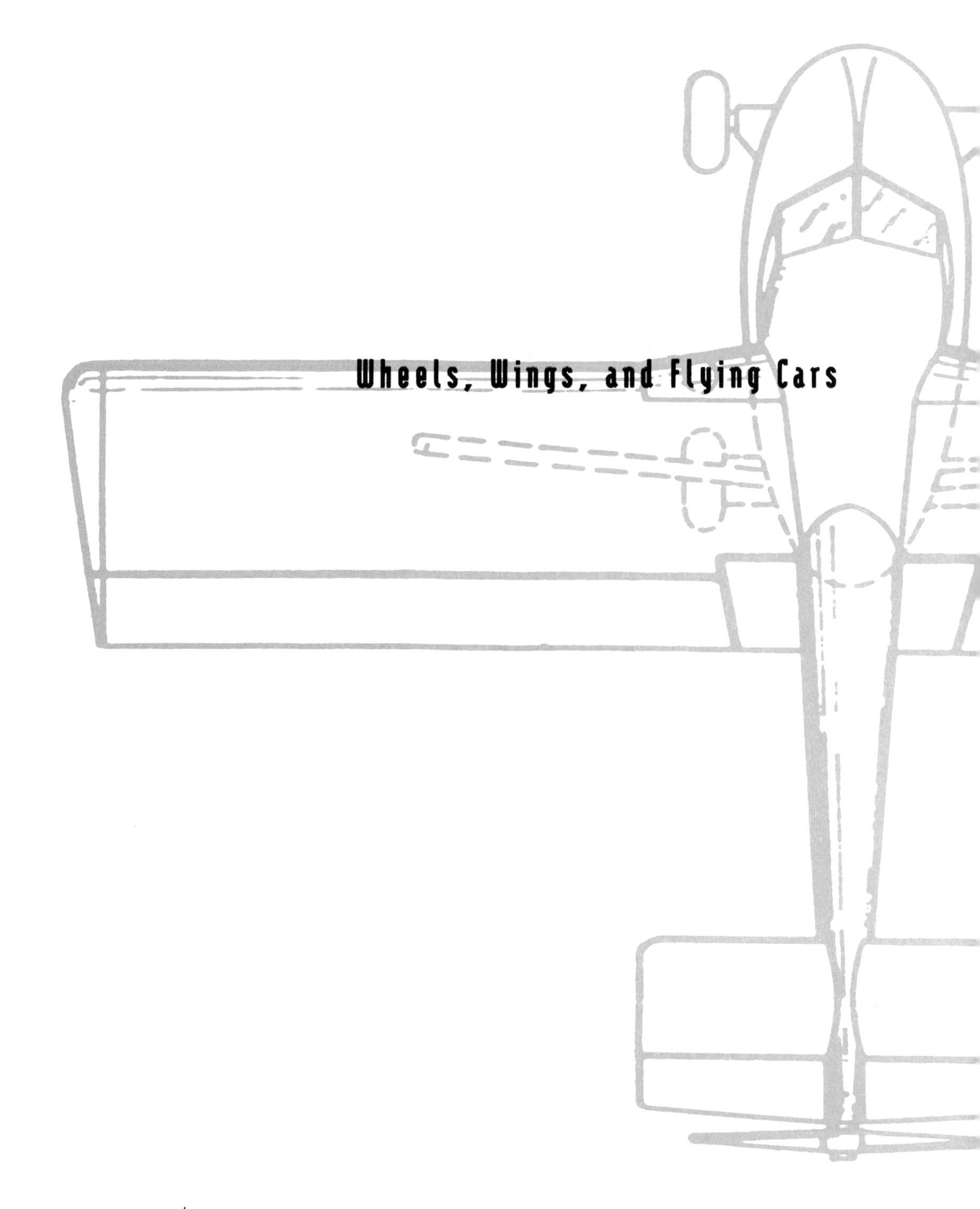

Wheels, Wings, and Flying Cars

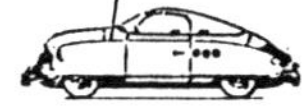

Saving the Miners in Dawson

1897–98

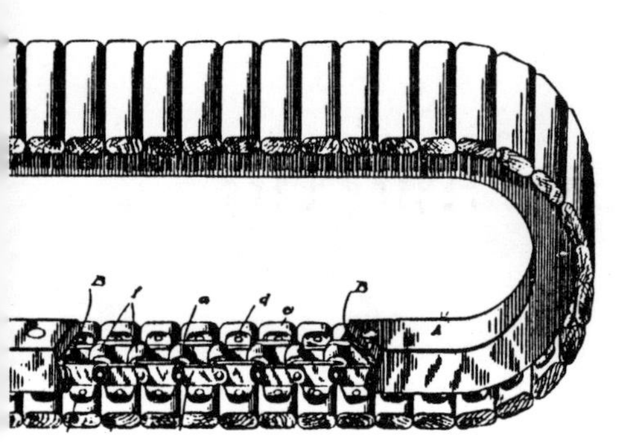

Otto Palmtag's tread design was one part of his "moving house," a steam-driven monster meant to conquer the ice and snow of the Klondike.

During the winter of 1897–98, when a number of Klondike miners were snowed in at the gold mines around Dawson, the crisis inspired a number of Northwest visionaries to create farfetched schemes designed to rescue the trapped men.

Eugene resident George Miller (brother of journalist/poet Joaquin Miller) proposed a gigantic hot stove, a sort of mountain steamroller, that could be dragged over the 700 miles of snow between Skagway and Dawson. The idea was to melt the snow and let it refreeze into ice, and "the long and weary way to the Klondike country will become a continuous series of skating rinks and toboggan slides, over which dogs and sledges can jog merrily along in the brilliant glow of the aurora borealis, carrying food and drink to the suffering miners, and return laden with gold." Miller proposed the scheme to the Pacific Coast Steamship Company, offering $500 of his own money to help develop the plan. They declined.

Around the same time that George Miller proposed his giant steamroller, D. H. Sterns of Portland invented an amphibious steam-driven toboggan, the 20-foot-long Sterns Terrain Flattener, which would "make a trip to the North Pole a comparative pleasure excursion." It featured wheels for land use, fins and centerboard and paddles for the water, treads for snow, and a winch and ropes for hauling itself up cliffs. Alas, it was never built. Otto Palmtag, of New Whatcom, Washington, similarly proposed a "moving house," a steam-driven sled big enough to carry a 7' × 24' house. It was to weigh three and a half tons, sink only six inches in the snow, move by means of wooden treads, and carry eight passengers and a crew of two. Palmtag formed a company to promote the machine, but before any stock could be issued the ice had melted at Dawson and the crisis was over.

Perhaps the most amazing proposal of all, however, came from Jefferson Dorsett of Tacoma, who suggested building hot air balloons big enough to carry a ton of supplies. These were to be carried, unfilled, to the base of a mountain pass and then inflated. Hawsers would connect balloon to cargo to land, and "the unencumbered argonaut can gaily climb the pass, towing his gear and duffle as a boy might fly a kite."

Gas Station

1907

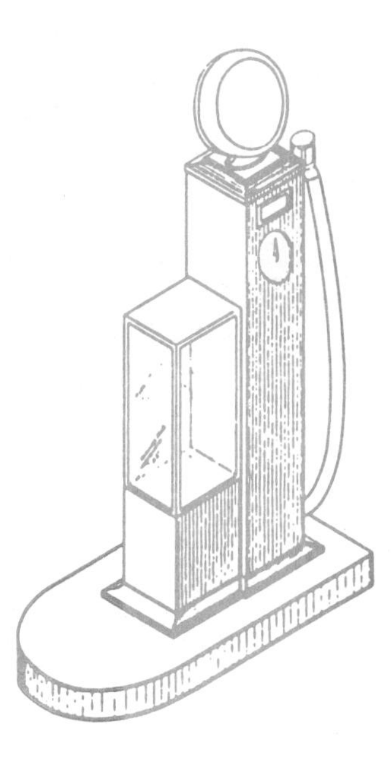

Motoring caught on early in the Northwest: by 1913, there were 19,497 automobiles registered in the state of Washington alone. The first chapter of the Good Roads Association, a group that lobbied for more and better roads, was organized by businessman/philanthropist Sam Hill in 1908, the same year that the Model T was introduced, and Washington was the second state to lay a road of Portland cement (after Michigan). In these early days, most drivers bought gasoline by purchasing a wooden box containing two five-gallon cans from a general store or livery stable. These shops usually kept their bulk store of gas supplies in portable galvanized tanks protected by small wooden houses built around them. The term "filling station" was derived from these wooden stands, and was coined by a gasoline-distribution firm called the Bowser Company of Fort Wayne, Indiana, which registered the trademark in the 1890s.

John McLean, the head of sales in Washington State for Standard Oil (Chevron) of California, changed the old system of selling gasoline irrevocably in 1907. By building what is generally considered to be the world's first gas station, McLean started a trend that was to have worldwide implications. He bought a piece of land adjacent to Standard Oil's main regional depot, at the corner of Western Avenue and Holgate Street on Seattle's waterfront. He and engineer Henry Harris then rigged a feed line leading from their main storage tank (housed at what was then called a "bulk station") to a 30-gallon, six-foot-high galvanized tank equipped with a glass gauge and a valve-controlled dispensing hose. This was

The pump heard 'round the world.

the first time that a piece of property, equipped with special machinery, was specifically set aside for the job of dispensing gas directly into cars, and the rest is petroleum history.

A plaque honoring the world's first gas station can be seen at Seattle's Waterfront Park.

Beaver State Motor Company

1912

The only automobile ever manufactured on a commercial basis in the Pacific Northwest was produced by the Beaver State Motor Company, founded in 1912 in Gresham, Oregon. The plan was to produce two models, the Portland and the Pacific, with aluminum chassis, oak frames, and six-cylinder, 45-horsepower engines. All the parts for the Beaver State cars were to be made at the Gresham factory except for a special Daimler-Lanchester worm-drive gear imported from England. Gresham was touted as the future "Wheel City of the West," and great things were predicted by the officers of the company. Unfortunately, the design of their engine so closely resembled that of the Overland Motor Company in Toledo, Ohio, that Beaver State was successfully sued for copyright infringement.

Two, three, or four cars were actually produced — the records vary — before the company gave up and turned to producing materials such as railroad car wheels, sewer pipe, and other materials. By 1924, the company was bankrupt. The last known Beaver State car was purportedly acquired by a Portland bankruptcy lawyer named Coan, whose children drove it until the tires gave out and in whose backyard it remained until 1929, when it was destroyed in a bonfire. According to legend, one month later a representative of the Harvey Firestone Company appeared in Coan's office and offered him $5,000 for it.

Early Land/Sea Vehicle

1914

Like many of his compatriots in the wild and woolly world of barnstorming aviation, Vern Gorst, a pioneer pilot from Coos Bay, Oregon, and an important figure in the early days of air mail and air cargo transport in the Pacific Northwest, was a born tinkerer. With another pilot named Charles King, he built an experimental boat with a Curtiss OX2 airplane engine mounted on the rear deck and its propeller facing backwards like the "airboats" used in the Florida Everglades. It went so fast that, in 1914, Gorst built a more sophisticated version, a land-and-water machine that Edgar McDaniel of the *North Bend News* dubbed "the Amphibian." It was made from an old Hupmobile chassis, with pontoons on either side and a platform for passengers or freight where the original Hupmobile engine went. The Amphibian was capable of doing 70 miles per hour on the beach, but only 15 in the water. Gorst began a regular passenger and freight service known as the "Fast Freight to Florence," which went from North Bend across the bay to Jarvis Landing, over the sand dunes, up the ocean beach, across 10 Mile Creek to the Umpqua River, out through the surf, up the river to Gardiner, and by road and beach to Florence — a distance of about 50 miles.

Automatic Transmission

1924

Several experimental automatic transmissions were tested in America and Europe during the early days of motoring, including the Sturtevant of 1910, the Constantinesco and DeLavaud transmissions of the early 1920s, and a two-speed planetary transmission patented by L. Renault in 1922. One of the many inventors who tried their hands at automatic transmissions was Joseph Fay of Seattle, who developed a self-shifting, variable-speed transmission in the mid-1920s —

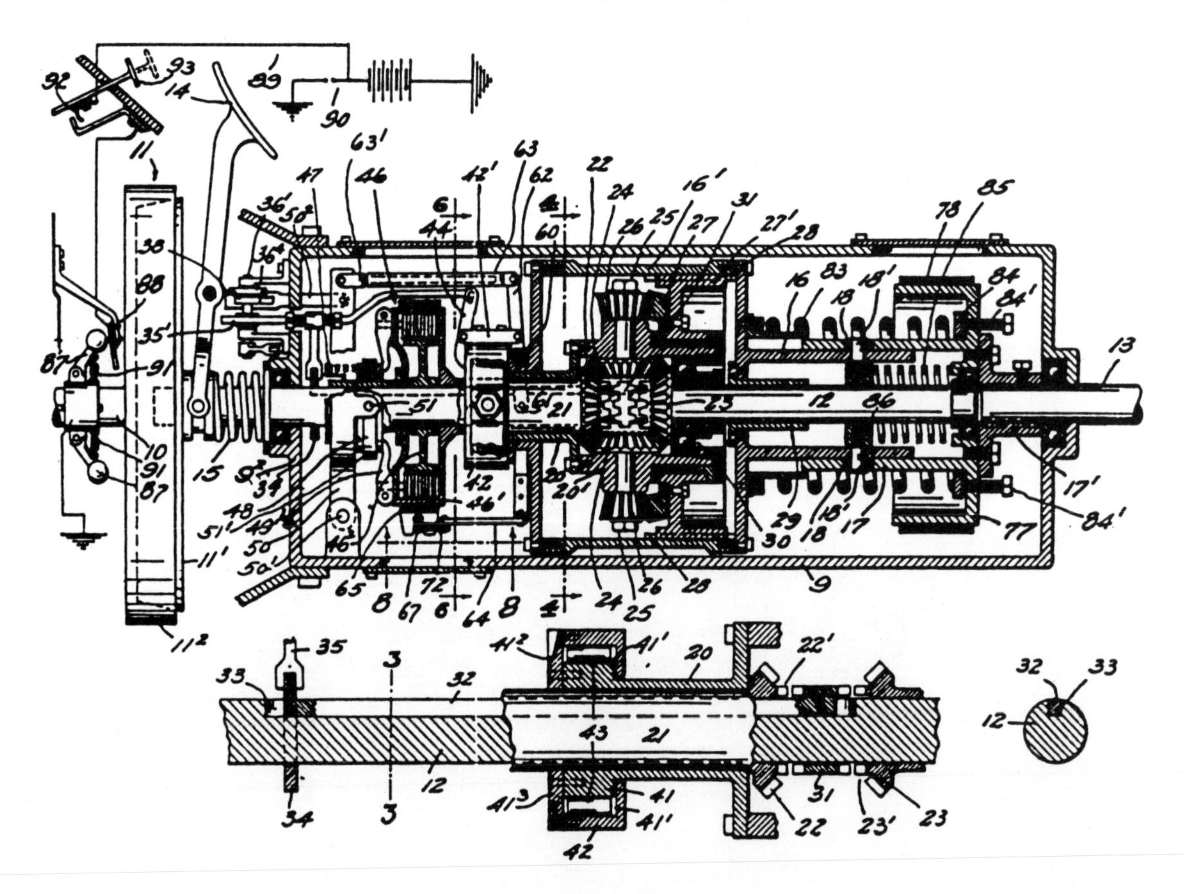

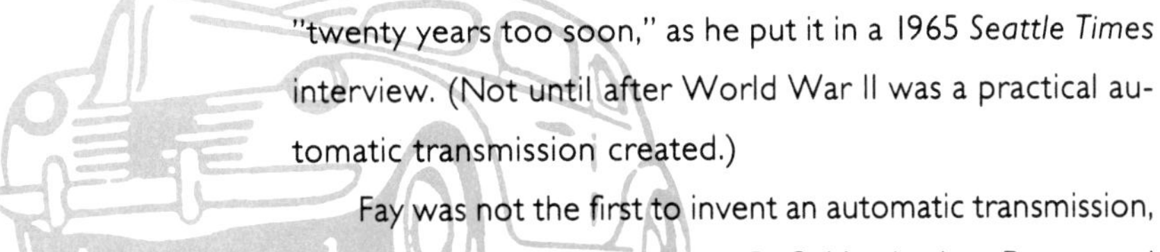

"twenty years too soon," as he put it in a 1965 *Seattle Times* interview. (Not until after World War II was a practical automatic transmission created.)

Fay was not the first to invent an automatic transmission, but he was certainly a pioneer. R. G. Harris, then Patent and Trademark Counsel for the Ford Motor Company, stated in a letter written to the Seattle Historical Society in 1964 that "it would be impossible to say who invented the first one.... [However] it is safe to say that Mr. Fay's device represents an early form of automatic transmission responsive to torque demands and, according to his account, would have been one of the few in actual operation in the early '20s."

Fay, the owner-operator of a Seattle moving company, was inspired to work on his automatic transmission because the truck drivers in his employ were regularly damaging his vehicles through careless driving. He spent eight years and $30,000 developing his machinery, which comprised a differential planetary gear with two clutches and a brake, connected to a torque (or load) governor that automatically shifted the mechanism into a lower gear when the grade became steep and the engine's running labored. Fay installed his transmission in a "Whisky Six" Studebaker, which he drove several times to Detroit in an unsuccessful attempt to interest manufacturers. The General Motors Corporation, following a 1924 demonstration of the transmission in Detroit, filed a report that stated: "The transmission is not now in such a state of development as to justify any immediate action on our part, for the reason that, aside from any question as to the performance of the device, the great size and weight makes its use impractical in its present form. The case is about 12 inches in diameter, by 30 inches long, and the weight is 200 lbs." Fay's transmission had one especially

Joseph Fay's automatic transmission was too heavy and expensive to find favor with car makers or buyers.

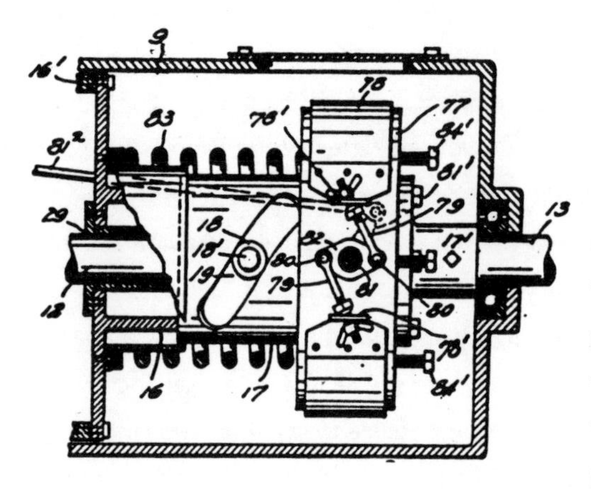

interesting feature: it was possible to jockey a car into a tight parallel parking space by rocking the car alternately between "go ahead" and "reverse" modes, thereby "fudging" it sideways.

The Museum of History and Industry in Seattle has a Fay transmission in its collection, though not on permanent display.

Bardahl Oil Additive

1939

Bardahl automotive oil additive, a top-secret formulation that is designed to improve engine performance, was invented by Ole Bardahl, who was born in Trondheim, Norway. Bardahl worked as a warehouseman in Norway until he immigrated to the United States in 1922. According to legend, he arrived in Ballard, the traditional Scandinavian community of Seattle, with $32 in his pocket and no knowledge of English. He worked in a sawmill, taught himself English by listening to downtown street preachers lecture about familiar chapters of the Bible, and took night classes to complete a high school diploma. Later he became a floor layer, then branched out into his own highly successful floorlaying and housebuilding business.

He recalled in 1979: "I was always interested in machinery, and in the thirties it became apparent that our car engines were demanding more of lubricants than ordinary oil could provide and that other additive oils would soon be necessary." The only lubricant available at the time was graphite, which was not effective for high-combustion engines. Working in his basement, Bardahl spent several years developing a formula for an additive that could be suspended in oil but would bond to the metal surfaces of engine walls so that it would not drain away. The additive was designed to increase the film strength of the motor oil, delaying or stopping its breakdown from friction heat, and thus reduce engine wear by up to 40%.

There is still some debate in the automotive industry as to its effectiveness, but the additive quickly found popular favor and remains popular. In 1939, Bardahl opened a small factory in Ballard to manufacture his product. Sales for his first year of operation were $188.53, but within eight years they had risen to $200,000; today the company's sales are in the millions, with manufacturing and distribution worldwide.

A series of advertisements that debuted in 1953 and spoofed the deadpan "Dragnet" TV police show helped boost Bardahl's popularity tremendously. These ads, over which "Dragnet" producer Jack Webb once threatened to sue, featured a trio of villains named Blacky Carbon, Sticky Valves, and Gummy Rings. Another reason for Bardahl's commercial success was his highly visible sponsorship of all types of racing vehicles, from motorcycles and snowmobiles to Indianapolis 500 cars. Perhaps the most famous of these racers was the bright-green *Miss Bardahl* hydroplane, which was world champion in 1961.

Open Deck Grating for Bridges

1933

The first use of open steel-mesh deck grating on a bridge occurred in Seattle in 1933 on the University Bridge. The grating, a matrix of modular steel pieces that could be riveted together in a variety of configurations, had been patented the year before by Walter F. Irving, president of the Irving Subway Grating Company of Long Island, New York. (Irving, an 1896 graduate of Rensselaer Polytechnic Institute, was a former steel salesman and ornamental-iron manufacturer who had made his initial reputation as an inventor with an improved grating for New York City subway vents. The vent featured linchpins that couldn't be stolen or vandalized, thus eliminating a major headache for New York's city engineers.)

Steel-mesh grating was a tremendous improvement over timber, the predominant material in use up to that time for bascule bridges — drawbridges (like the University Bridge) that balance like teeter-totters when they rise to let ships through. Steel grating was easier to raise and lower because of its high strength-to-weight ratio and ease of drainage. Horses and cattle could cross it safely, obviously more of a consideration in 1933 than now. It presented few problems in high winds, unlike solid-plate bridges, and proved to be more durable than wood. Previously, Seattle bridge engineers had to completely replace the wood on the University Bridge every 10 to 20 years. There was no need to do anything more than routine maintenance on the 1933 grating until it was replaced in 1990, over 50 years after its installation.

The characteristic buzzing made when a car crosses a bridge on open steel-mesh grating was first heard in Seattle in 1933.

John A. Dunford, Seattle's chief bridge engineer in the 1930s and 1940s, was so impressed with steel-mesh grating that he subsequently oversaw its installation on the Fremont, Ballard, and Mercer Island bridges. (The original Tacoma Narrows Bridge, "Galloping Gertie," was not equipped with the Irving grating; after it fell during a windstorm in 1940 — an event recorded on a famous home movie — the grating was used to break up the air flow and prevent another collapse.) It has been estimated that 70% to 80% of all the drawbridges in the United States now employ steel-mesh grating, although only 20% to 30% of new drawbridge construction uses it because of the several alternatives, such as solid plates or grating that has been partially filled in with concrete along tire routes, that can be used instead.

Although steel grating has been widely advertised as being safer in wet weather than other types of surfacing, the engineering community is not of one opinion on the issue. Recent refinements, such as a serrated top, have significantly improved its safety quotient in wet weather, and all agree that it is a radical improvement over wood. Seattle City records indicate that the University Bridge averaged 182 accidents and 6 deaths annually before the grating was installed, but no deaths or accidents due to slipperiness have occurred since the installation.

Phibian Fluorocarbon-Powered Car

1930s

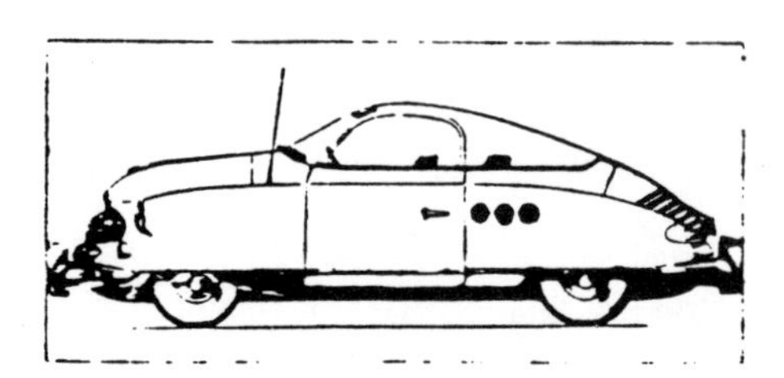

Greek emigrant Constantinos Vlachos was known around his adopted hometown of Grand Coulee, where he had settled in the 1930s, as both an inventor and a one-man protest rally. In his later years, he'd sit on a chair in front of the tiny, cluttered "museum" that housed his memorabilia, talking with passersby and honking a loud bulb horn every time an automobile drove past. The horn was his way of protesting America's dependence on oil-powered vehicles, and of expressing his dismay at the cold shoulder the world had given his own invention, the Phibian, a car that ran on fluorocarbons instead of gasoline. (According to those who knew him, it was also a way for the outspoken Vlachos to attract conversational partners.)

Vlachos began work on a Phibian prototype in 1935 and was still working on it at least as late as 1949, when he demonstrated the car at Portland State University. The Phibian featured a streamlined design with a clear plastic top. Each wheel was powered by its own baseball-sized, eight-piston hydraulic engine. Vlachos battled most of his life for the recognition he felt the Phibian deserved, bombarding government authorities with a nonstop stream of letters and position papers. His personal stationery and business cards featured the slogan "68 Years To Convince The World," and an autobiographical statement he distributed begins, "No City, State or Federal Government has stopped my work. I have fought for years for clean Government. My record speaks for itself. I pray to God to punish all of them and I am not sorry."

In particular, he wanted the federal government to allocate $500,000,000 for the production of 25,000,000 Phibians, at a manufacturing cost of $20 each and a retail price of "$50 complete." He claimed that in 1951 he was offered a million dollars for the rights to the car by "three large oil companies," but turned them down because he feared losing control over the invention. (In light of current knowledge about the detrimental effects of fluorocarbons on the environment, it is perhaps just as well that the idea never garnered public favor.)

Vlachos also designed the Tri-Phibian, a vehicle that traveled on land, sea, and air. He built a prototype in a Pelham, New York, workshop in 1935 and drove it to Washington, D.C., to demonstrate it in front of a group of government officials that allegedly included President Franklin D. Roosevelt. According to the inventor, the toll collectors at New York's Holland Tunnel, which connects the city with New Jersey, refused to let the Tri-Phibian through the toll gates, so Vlachos backed up, turned on the gyrocopter blades, and "flew over the Hudson instead of under it." The Tri-Phibian blew up during the demonstration in D.C. Vlachos was severely burned and had to scuttle further development plans. For the rest of his life he remained convinced that unscrupulous agents of the government and the big automobile manufacturers had sabotaged his invention.

Vlachos devoted many of his later years to creating elaborate mother-of-pearl representations of the Last Supper

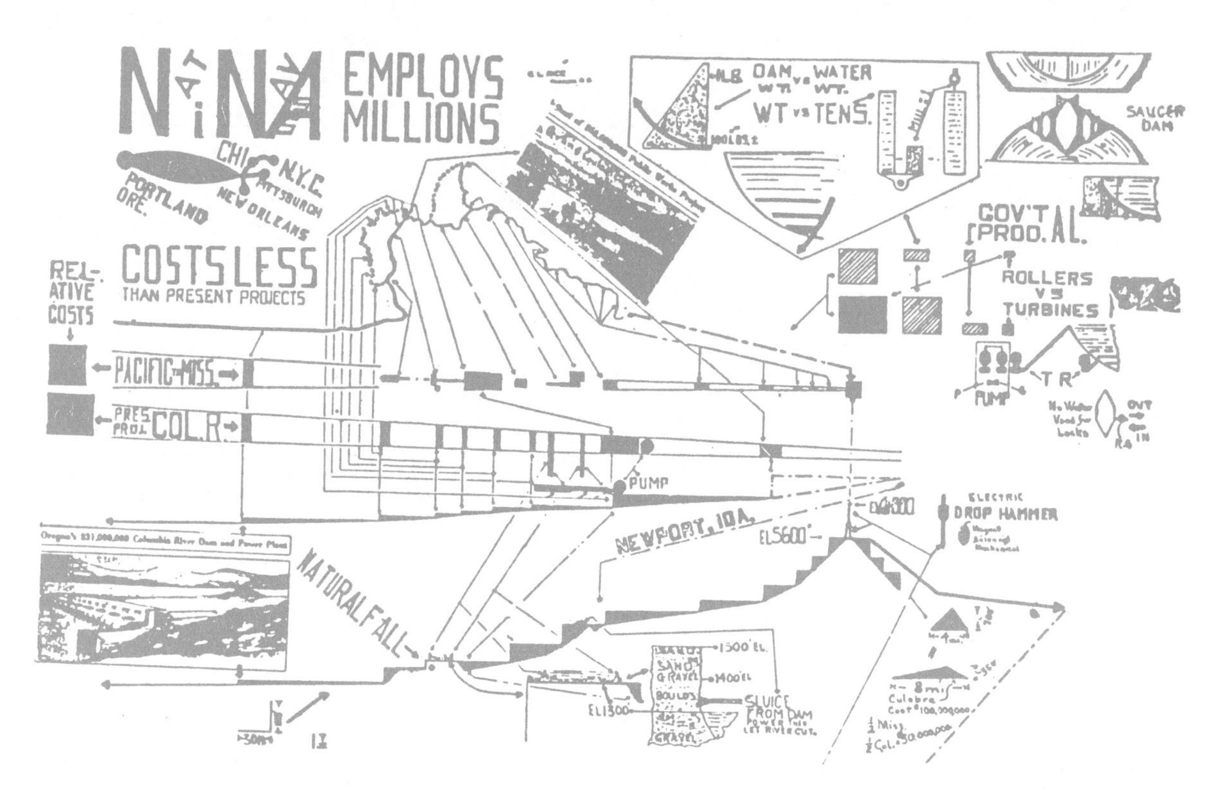

In addition to the Phibian, Vlachos's designs included a machine gun that fired sleep-inducing pellets, a greaseless pump that he said could make 22,000 revolutions per minute, a pilotless airplane designed to fly at 100,000 miles per hour, and the irrigation/dam system shown here.

and Crucifixion. He also concentrated on his museum, which contained, according to an autobiographical statement,

> millions of books and magazines, lots of historical pictures, different stages of life, Tri-Phibian pictures... a rough sketch of the St. Lawrence Waterway System. Pictures of old cars, international defense pictures, Inventor city...Expensive 100-year old magazines, old radios, old desks, new condition antiques and stainless steel...Any engineering book, I have, university professors should learn my work is 100 years ahead of time. Health lectures...old purses, bronze pictures from Greece and Athens...books on the Russian Revolution...the #1 telescope that reaches 48 million miles, the #2 telescope reaches 48 to 96 million miles...[and] three types of postcards...for sale at 50 cents each.

Air-Cushion Vehicle

1957

The story of Walter A. Crowley, a retired engineer in Oak Harbor, Washington, is one of opportunity just barely missed. Although he has received little recognition, Crowley was a pioneer in the development of the ground-effect machine, also known as the air-cushion vehicle. At more or less the same time that Crowley was working on his design, a British engineer and boat architect named Christopher Cockerell

was working on a similar idea; due to a succession of circumstances, Cockerell's plan, now known as the Hovercraft, was the one that gained acceptance.

In the mid-1950s Crowley worked as an engineer for General Motors and Chrysler's missile division. In his spare time, he tinkered with motor cars, robot puppets, and other toys for his children. He also designed and built his own home, which featured many original built-in gadgets, not all of them foolproof. More than once, the electric garage door opener "delivered a healthy jolt of electricity whenever it rained," according to the inventor's son, Seattle political activist/writer/commentator Walt Crowley, and the TV always managed to fall out of its built-in wall recess at crucial moments. The bathroom was equipped with a sliding door that opened by means of a push-button; during one New Year's Eve party it suddenly and uncontrollably began opening and closing of its own accord, exposing to public view a mortified female guest in a state of undress. (In an unpublished memoir, the junior Crowley writes: "My father's later musings about installing an electric toilet drew an icy reproach from my mother.")

Crowley's experiments with air-cushion vehicles were inspired by a common kitchen phenomenon. A coffee cup taken from a hot-water rinse and placed upside down on a slanted drain board will rise slightly off the surface and slide down the drain; momentary expansion of the hot air inside the cup lets it skid on an almost friction-free air cushion. In 1956 Crowley took this by-product of the Bernoulli Principle — known to airplane pilots as "ground effect," a momentary uplift experienced when landing an airplane — and designed experiments to reduce the surface friction between a vehicle and the surface it travels over.

He began by attaching the outward-blowing end of a vacuum cleaner to the tops (bottoms, really) of a variety of upside-down buckets, lampshades, and other containers. By blasting air into them downwards, he could cause them to rise slightly and float on a cushion of air. He was soon experimenting with larger "vehicles," including a full-size model built in the fall of 1957 in the basement of his home in Royal Oak, Michigan. This model consisted of a large canvas cone stretched over an aluminum frame. Inside the cone was an electric motor powering a horizontal propeller, and resting on top was a platform large enough for one person. Crowley Jr. writes, "I remember when my father first switched on his creation. The blast of air scattered laundry and dusty bric-a-brac and lifted the machine an inch above the floor. Looking like an overgrown green lampshade, the thing torqued and glided freely around the basement. My father gingerly mounted the platform, but his adult weight was too much for the motor. So, at the age of ten, I was recruited to become the first 'man' to fly an air-cushion vehicle. The precise date of this indoor Kitty Hawk has been lost, but I remember that we were giving demonstrations to the relatives by Thanksgiving 1957."

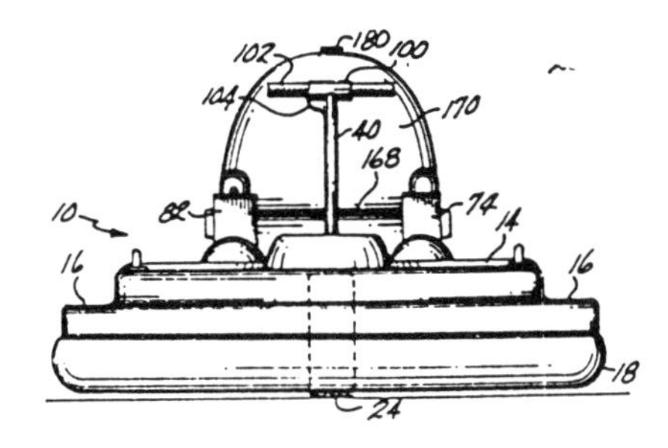

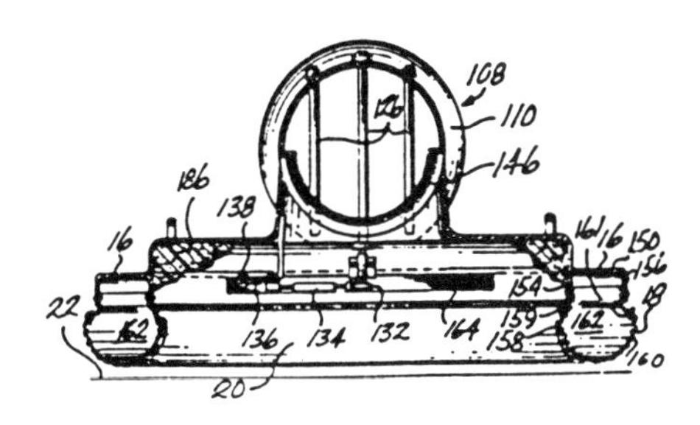

Walter A. Crowley's early "air-cushioned and ground-engaging vehicle" was aced out of the running by the British-designed Hovercraft.

That winter, Crowley produced a new model, capable of supporting an adult, that incorporated a gas-powered engine and a 9' × 16' oval cone. The cockpit of this model had a seat, windshield, dashboard, and steering wheel; thrust and direction were provided by a second engine mounted ahead of the pilot. When the inventor first took it out for a spin in the spring of 1958, the neighbors called the police, Crowley's son reports, about "some nut buzzing around his backyard in a flying saucer," but this didn't keep him from testing his machine later on local streets and in golf courses, parking lots, and other public spaces.

Crowley applied for a patent on the design in 1958, but he ran into the all-too-common problem of indifference on the part of corporations that might have helped to mass-produce it. (In England, Cockerell was experiencing the opposite problem: too much attention. His project had been deemed top secret by the government, and he found himself

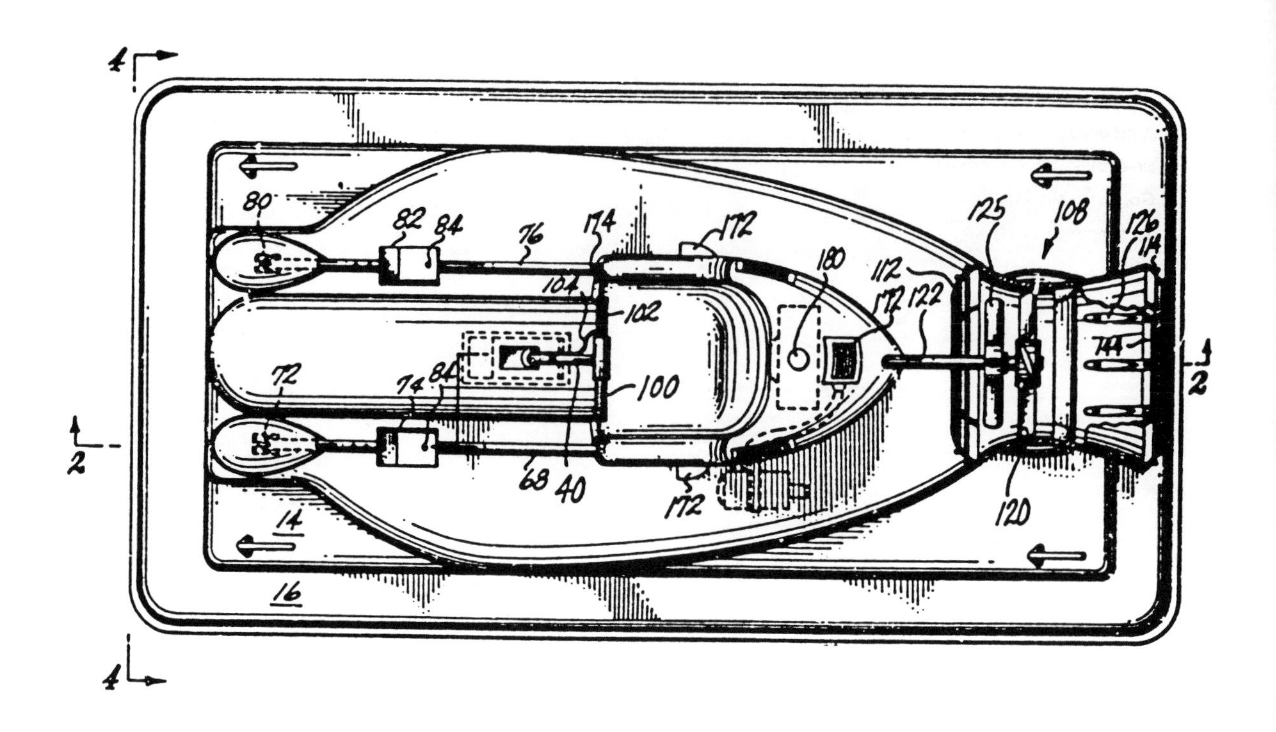

mired for several years in red tape.) When Crowley found backers in Washington, D.C., later that year who agreed to buy the manufacturing rights to the device, he moved his family to Bethesda, Maryland, and a period of considerable publicity followed, including a demonstration for the House Science and Space Committee.

In early 1959, Crowley landed a contract to produce a prototype vehicle for the U.S. Marine Corps: an all-aluminum, 30' × 24' craft with a 270-horsepower engine, capable of carrying a two-ton payload on water or land. The prototype was completed late that year, but Crowley's business associates suddenly declared bankruptcy and disappeared with all the company's assets, including the Marine prototype. In England, meanwhile, Cockerell had overcome his initial bureaucratic tangles and was enjoying considerable success. The first Hovercraft crossed the English Channel in 1959, and had substantial backing from both government and private

A modification of Crowley's air-cushion car was the Aero-Go, an air-bearing pad for lifting extremely heavy objects.

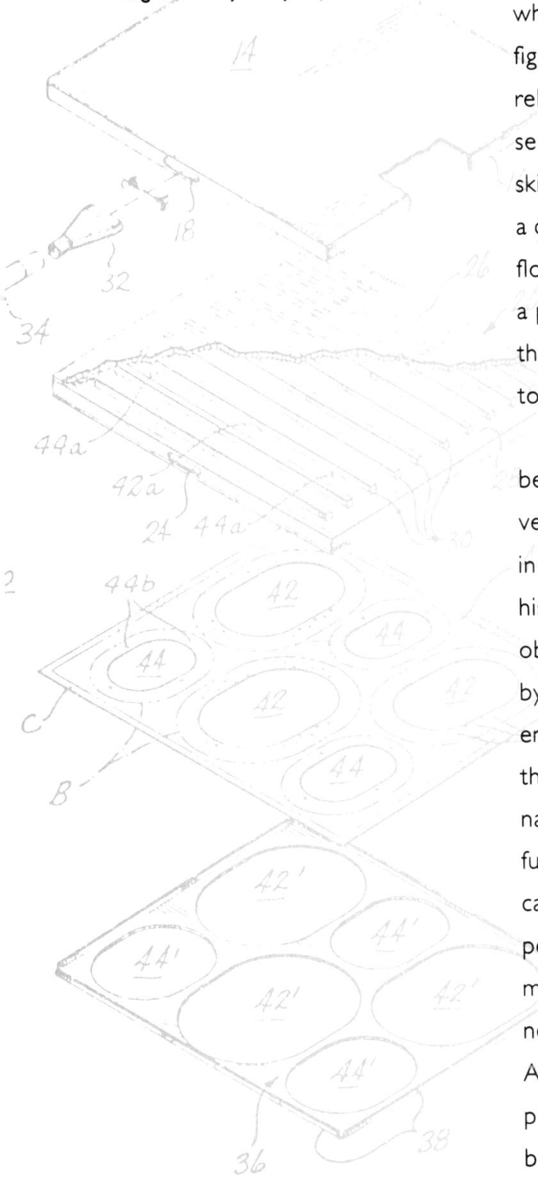

industry; in time it was accepted as the world standard.

The main difference between Crowley's and Cockerell's designs was that Crowley preferred a plenum chamber, while Cockerell favored an annular or peripheral jet configuration. A plenum-chamber jet operates by forcing air at relatively low pressure uniformly downward in what is essentially an open box. The air is then trapped by a flexible skirt at the bottom of the box and slowly leaks out, forming a cushion. An annular jet works by directing a high-pressure flow from the inlet at the center top of the mechanism over a perpendicular interior wall. This stream of air is forced to the edge of the chamber and is thrust downward and inward toward the ground.

In 1961 the family came to Seattle, where Crowley had been recruited by the Boeing Company and where he developed an air-bearing pad (for which a patent was awarded in 1968) for moving heavy equipment. It is an adaptation of his original vehicle design that permits the lifting of inanimate objects and is controlled by an outside operator instead of by someone sitting on top of it. As is required of Boeing employees, Crowley signed over the manufacturing rights for the air-bearing pad to the company; under the subsidiary name AERO-GO, it has been one of Boeing's most successful nonaircraft products. The standard AERO-GO machinery can be plugged into ordinary shop air operating at 90 pounds per square inch, and the low-pressure plenum configuration means that once the pad is inflated, very little pressure is needed because almost no air escapes from the skirt. The AERO-GO's uses include moving stadium bleachers from place to place and sending heavy equipment along assembly lines. It is even used to move entire airplanes: an air-cushion pad is placed under each landing gear and the airplane can revolve in place. The AERO-GO uses a flexible skirt, designed by Crowley, to keep the lifted object balanced and highly maneuverable, even when floating a fraction of an inch off the ground. A similar flexible skirt is standard equipment today on virtually all Hovercraft and allows a smoother ride over the waves.

Boeing 727

1958–1962

The Boeing 727 had a wing-span of 108 feet and a length of over 133 feet.

The Boeing Company has been awarded so many patents, and has been responsible for so many sweeping innovations in so many branches of aeronautics, that it is impossible to choose one invention as being the most significant. To single out one individual as the overriding force in the company's achievements is also difficult, because Boeing's design methods favor the use of large committees and teams. But by looking at the development of one aircraft — the 727 short-to-medium–range jet transport — it is possible to glimpse the workings of the creative process on a large scale and to see the way in which complex pieces of airplane machinery are made.

The 727 was an amalgam of both tried-and-true design practices and untested innovations; it was also a commercial product carefully conceived to fill a specific market niche. In "Billion Dollar Battle," a book-length study of the plane, Harold Mansfield writes that the decision to create a small, short-range jet came in 1958, in the wake of the first tests of the 707. (The hugely successful 707 was the world's second swept-wing jet transport, using a design that had been perfected with the company's B-47 and B-52 jet bombers. It was also Boeing's sixth large-scale entry into commercial aircraft production, and one of the world's first quantity-production jet transports.)

Jet travel was becoming increasingly popular, but only Sud Aviation of France had successfully designed a small jet geared for brief (150- to 1,700-mile) journeys. Such a plane had to be able to handle relatively short runways — the benchmark was LaGuardia Airport's shortest runway, 4,980 feet — but the craft also had to make economic sense. Jet engines perform most efficiently, and therefore most cheaply, at high altitudes, but the fuel costs incurred in getting to cruising altitude generally favor large, long-range aircraft; all the fuel on a small plane, it was argued, would be used up simply in gaining altitude. Turboprop jets, such as the British Viscount and the Lockheed Electra, were compromises, combining as they did turbines and piston-driven propellers, but they offered less attractive long-range economy and a bumpier, less pleasant ride than did straight jets or prop-driven planes. The answer was a small but efficient jet.

Boeing financial analysts estimated the production cost of a craft in this category to be between $3.25 million and $3.5 million per plane. At this rate, Boeing would need to sell 200 just to break even, but if the unit price could be brought down to between $2 million and $2.5 million, the company could expect to sell 180 planes within five years and see a profit. Because of the project's extremely high development costs, estimated at upwards of $100 million, the risk would be much smaller if airlines would commit to buying substantial numbers of the new craft before production began. It was therefore crucial to keep the jets as inexpensive as possible.

As far back as 1956, Boeing's Preliminary Design Unit had been drawing sketches of a short-range jet. Some versions had two engines, some four, and some three. When John E. ("Jack") Steiner, a Boeing engineer who had worked on the 707, reentered the Preliminary Design Unit as assistant chief in mid-1957 and became head of the 727 planning group, 38 separate variations on the basic theme had already been tried. Similarities to aircraft being developed or introduced by the competition, such as the de Havilland Trident, Douglas DC-9, and Sud Caravelle, also had to be taken into consideration. For example:

•Number of engines. A four-engine 727 might be too large and crowd the midrange market for Boeing's larger 720. On the other hand, two-engine jets were falling out of favor for safety reasons. A three-engine design, new and untested, would require considerable development costs: the tail engine would need to be sized differently, and a long air intake passage and larger tail surfaces would be required. In its favor was a lower estimated operating cost: only 3/10 of 1% higher than that of a two-engine configuration of the same class. (A four-engine model would have been 6% higher.)

•Engine size and type. In the past, airplanes had always been developed around existing engines, but engineer George Schairer suggested writing up specifications for a new engine.

•Engine placement. Mounting engines on the aft portion of the fuselage reduced noise and made a more efficient wing. The main disadvantages were less passenger space and increased problems in balancing the plane on the ground. Wing-mounted engines created more passenger space and better weight distribution for a stable center of gravity. The disadvantages were increased noise and less efficiency in the wing.

In September 1959, configurations engineer Ken Plewes finalized Model 727-323A, the design that eventually became the one Boeing used. By the end of October, the plans had jelled: three custom-designed Pratt & Whitney JT8D engines mounted aft, one of them submerged in a low, T-shaped tail; a stairway that retracted into the rear of the plane; and a 148"-wide cabin. The landing gear retracted into bulges in the wing roots rather than the body, giving the airplane a larger aft cargo pit; the wheels were placed far enough back to keep the weight of the aft-mounted engines from unbalancing an empty plane on the ground. This design's direct operating costs were estimated to be below those of the Electra and DC-9, but its speed would be 50% greater than the Electra's and 20% greater than the Caravelle's. Overall production cost estimates had risen, however: by early 1960 the price had climbed to $3.8 million per 727. Boeing's sales department now figured they needed to sell a billion dollars' worth of airplanes.

In April 1960, the 727 was elevated to full project status. Steiner was appointed Chief Project Engineer, and under him were a number of key personnel, including Fred Maxam, systems; Milt Heinemann, interiors; Joe Sutter, aerodynamics; and Joe Miles, programs. Production began in early June on some 70,000 individually designed parts. By November, there were 500 engineers assigned just to the design aspect of the project, with 120 new ones added monthly. At its peak, 1,500 engineers were working on the project, with an average of 27% overtime. Boeing requisitioned six million square feet of plant area in Renton, Seattle, Auburn, and Wichita,

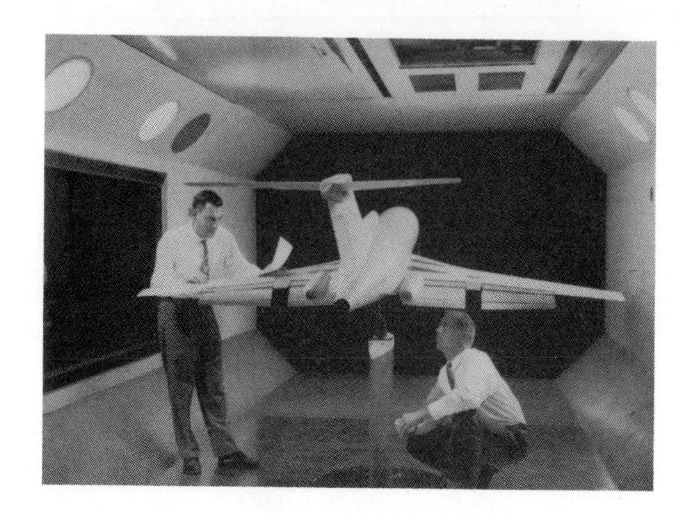

Kansas, and some 1,200 subcontractors and suppliers were signed up.

The entire 727 project, of course, would have gone nowhere if Boeing's client airlines had not indicated a need for a jet with its particular specifications, and they were consulted heavily during the design stage. In addition, Boeing engineers developed several new features on their own. Among the chief innovations that emerged were:

•A plan to keep variations to a minimum. With roughly 55,000 separate tool designs needed for assembly and 67,000 separate engineering drawings, it would have been impossible to completely customize planes for particular airlines. The "Omnibus Program" identified the areas of the aircraft (about 10% of the total) where choices could be made by individual airlines, then grouped them together.

•A set of strong "spoiler" air brakes to save time in descent and at the terminal and thus lower operating costs. Five minutes saved "block to block" meant a savings of a million dollars a year for a fleet of 15 to 20 planes.

•A wide-body configuration, important because of the growing popularity of tourist class.

•A 32-degree swept wing. This was a compromise between Boeing's favored 35-degree sweep, which required less wing weight so that more could be devoted to fuel tanks mounted inside the wing, and United's favored 30 degrees, a more familiar and more maneuverable configuration for its pilots.

•An Auxiliary Power Unit (APU), a small gas turbine capable of supplying electrical/pneumatic power and air conditioning for the airplane when its main engines were shut off. (The 727 was the first Boeing commercial jet that was so equipped.) The combination of the APU and the 727's self-contained aft boarding stairs made it essentially independent of ground power vehicles. (See the section on "D. B. Cooper" for an explanation of why the aft boarding stairs design was later modified.)

•An automatic compensator, devised by hydraulics expert Ed Pfafman as part of the flight controls, gave pilots the freedom to be unconcerned with changes in the plane's center of gravity as the passenger load shifted after each stop.

•An improved landing gear that swept back on a skewed axis, down and aft at the same time, with a dog-leg strut.

•A 130-foot full-size flight control testing structure, the "iron bird," with huge pneumatic cylinders that produced simulated maneuver conditions. (The flight control systems for the 727 had unique problems caused by the rear placement of the engines.) This structure was expensive, but saved money in the long run by obviating the need to make significant changes after in-flight testing.

•Cold and hot metal bonding instead of, or in addition to, riveting. This required extensive retooling, but meant a longer life for the body and wings of the plane.

•"Rubber tooling," which made key tools adjustable to accommodate future growth of the plane or new models in the line, as well as a computerized method of determining how to specialize tools. (These were well-known systems in the automobile field, but were relatively new to the airplane industry when Boeing introduced them.)

Ed Wells, Boeing's longtime chief engineer, played a major role in the 727's design.

Thousands of people played vital parts in designing and completing the 727. Among them were:

•Ed Wells, generally considered to have had more influence on the course of aeronautics than any other engineer in the industry's history, oversaw the project.

•Maynard Pennell, Chief Engineer for Boeing's Transport Division, a veteran of the B-47, B-52, and 707 projects, played a major role not only in the 727's development but in the overall development of jet aircraft for commercial use.

•Richard Tilson, head of Matériel, was in charge of ordering castings and forgings, which took the longest to make.

•Barney Storey, the project's chief tool and production engineer, oversaw the creation of "loft lines" — the shapes of general body contours — from which master models were made. These were then used to make moldings for metal dies for press-forming parts, and contours for jigs and fixtures to hold them while being assembled.

•Joe Sutter, chief aerodynamics engineer, and Bill Cook, chief of technical staff, improved the plane's lift capability with a triple-slotted trailing-edge flap system. Extra-high lift, necessary for short runways, was provided by putting slots at the front of the wings, which allowed high-speed air to come through from the underside of the wing to the wing's upper surface, where the wing's curvature set up a suction that provided lift. The design also permitted higher sweep and smaller wing area.

•Engineer Dick Weiland devised a system for the flap controls, gimbaling joints on carriages that traveled down an external track. Because of the triple-slotting design, the flap controls couldn't be mounted inside the wing. With this system, the tracks could run streamwise to the wind. It provided easy maintenance for ground crews while maintaining favorable lift characteristics.

Once the design and production schedules jelled, construction of the two preproduction test planes moved ahead quickly. The wing stub and wing were attached in August 1961. In September the body, nose gear, cockpit, and forward body were completed; the front and rear halves were joined; and soon after, the landing gear, stabilizer, and engine installation were completed. Finally, on November 27, 1962, at 12:05 P.M., a completed 727, painted yellow and brown, was rolled out for public inspection. Some 30 months had passed between go-ahead and finished product.

The airplane made its first test run, a taxi maneuver on the ground, on February 8, 1963. Within the next week four more tests were conducted. This was an extremely short schedule; in only a few days, the crew had completed tests that might normally take four to five months for a new plane of the 727's class. The plane functioned extremely well; the 450 hours it had spent in the "iron bird" paid off. The only major problem the crew encountered was an occasional surging or backfiring of the center engine, which was rectified with the addition of a vortex generator in the air intake. In fact, the 727's performance was better than predicted, because its improved drag gave it 10% better fuel mileage than had previously been assumed. This translated into an average of $2 million saved on each $4.2 million airplane over a 10-year period.

Boeing took its as-yet-uncertificated 727 on a world tour in the summer of 1963, culminating at the Paris Air Show the following spring. This was a bold move, since the plane's home base, extra parts and tools, and pool of engineers were far away. However, since the plane needed 150 to 200 air hours of performance and reliability testing for the FAA anyway, and since Boeing was eager to preempt de Havilland (rumored to be preparing the Trident for the air show), Boeing seized the opportunity for free publicity. The plane traveled 76,000 miles on 139 separate flights and visited 26 countries. It encountered no major problems.

Meanwhile, static fatigue testing of the second preproduction plane was completed on the day before Christmas, 1963. The metal finally split at 110% of the design strength, much better than expected, which meant that a higher payload would be possible. Just before Christmas, the airplane received final FAA certification. The first model put into service was a 727-100 for Eastern Airlines, which began operation in February 1964. The 1963 "basic flyaway" price for this first plane was $4.3 million.

Variations on the 727-100 included the -100C, a model that could be configured to carry either passengers or up to 46,000 pounds of cargo. The -100QC, a quick-change convertible, could be altered using palletized seats and galleys and pallet rollers. At the end of a typical passenger flight, the seats and galleys detached from locks and rolled out through the cargo door into a storage van. The cargo pallets were then rolled directly into the plane and locked in place, with the entire operation requiring less than an hour.

The 727-200 series (the first of which was delivered to Northeast Airlines in December 1967) expanded the -100's capacity from 94–131 passengers to 134–189 passengers. The -200, a medium-range plane with a range of over 1,500 miles, was essentially the 727 with a longer body. The last 727, a -200F freighter produced for Federal Express, was delivered in September 1984.

Airplane Hijack for Money

1971

On a cold autumn day in 1971, a man calling himself Dan Cooper boarded a flight from Portland to Seattle and executed the world's first successful airplane hijacking-as-extortion. (Although he called himself "Dan Cooper," he became "D. B. Cooper" when an FBI clerk misidentified him as such to a UPI reporter.) When "Cooper" bailed out over southwest Washington with $200,000 in cash, he became an instant folk hero. Despite one of the most highly publicized manhunts in FBI history, he has never been apprehended — assuming, that is, that he survived.

Cooper was not just a folk-culture phenomenon. His crime led to a change in the design of subsequent models of the Boeing 727 — a locking device known as the "Cooper Vane" now prevents the airplane's stairs from dropping while in flight. The various increased security measures that are now standard in airports, particularly the X-ray procedures made mandatory by the FAA in February 1972, are also, at least partly, traceable to him.

Cooper bought a one-way ticket from Portland to Seattle just before the departure of Northwest Airlines flight 305 on the afternoon of November 24, 1971. He paid the $20 fare in cash. Nothing about him aroused suspicion; his appearance and manner were typical of the many businessmen who frequented the commuter flight. His only luggage was an attaché case. He chain-smoked, and he ordered a bourbon and water from the stewardess as the airplane taxied onto the Portland runway.

He also handed her a note that said he had a bomb in

his attaché case, which he opened to reveal an impressive-looking tangle of wires and sticks. Using the stewardess as a go-between, he ordered the captain to fly to Seattle as scheduled, where the passengers were released and the plane was refueled. He demanded, and got, four parachutes — two front, two back — and $200,000 in cash. When the plane was ready for takeoff in Seattle, he stayed in the rear of the craft with one stewardess, Tina Mucklow, and instructed the pilot to fly toward Mexico City, by way of Reno for refueling. He gave specific instructions to keep the airplane's flaps and landing gear down. Shortly after takeoff, he told Mucklow to go forward to the cockpit and not come out. Somewhere over southwestern Washington, with the 727's flaps and landing gear still lowered, the pilots saw the aft stair warning light come on, a signal that the stairs had been lowered. A few minutes later they felt oscillations in the airplane's atmosphere. Cooper did not respond to repeated attempts at communication via intercom. When the airplane landed in Reno and was searched, he was gone; the vibrations had signaled his bailout.

Cooper was obviously familiar with the layout of the 727. He knew that keeping the airplane's flaps and wheels down would slow the plane sufficiently so that he could bail out. He knew the stairs could be lowered while in flight. He knew the air traffic patterns and procedures around Seattle. He apparently timed his departure to land in the rugged country around Mount Adams. The CIA had used the 727's in-flight stair-dropping capability to drop large packages in Vietnam. Was Cooper a former intelligence officer? Was he a former Boeing engineer? Did he have a vehicle waiting for him? Was the bomb real? Did he survive his fall? Many alleged sightings of Cooper have been reported since his escape, but since only a small portion of the marked money he took has been recovered — a few packs were found by a young boy on the banks of the Columbia River in 1980 — it is probable that he did not survive.

Flying Saucers

1947

The sighting that ignited world interest in unidentified flying objects (UFOs), the incident that gave rise to the term "flying saucer," occurred above the Cascade mountain range on June 24, 1947. As Kenneth Arnold, a private pilot from Meridian, Idaho, flew near Mount Rainier, he saw nine swiftly moving, crescent-shaped objects, each "nearly the size of a DC-4," about 25 miles from his plane. Arnold, the founder of a Boise-based fire equipment company, recalled that "they flew as if they were linked together, swerving in and out of the high mountain peaks with flipping, erratic movements...like a saucer would if you skipped it across the water." Arnold reported his sighting to Noland Skiff, an editor at the *East Oregonian* in Pendleton, who wrote a story about it that included the term "flying saucer."

The *East Oregonian* story was picked up by news services worldwide and touched off a storm of UFO sightings. Within weeks, flying saucer reports had poured into news and governmental agencies from every state in the union and from several other countries. Others reported sighting non–saucer-shaped UFOs, including objects resembling ice cream cones, hubcaps, balls of fire, and cigars with lighted windows and blue flames along their underbellies. Although many who sighted them were convinced that the strange sky objects were spaceships from another world, newspaper and radio commentator Walter Winchell claimed otherwise: he said he had inside information that proved they were actually spy planes from Russia. (Andrei Gromyko, then a senior Soviet diplomat, declined to admit responsibility, although he did speculate that the culprit might be an overenthusiastic Soviet discus thrower.)

Arnold said in 1983 that he saw UFOs on seven different occasions after that initial sighting, the largest grouping being one of 23 or 24 vessels, all saucer-shaped, with a center spot that pulsated "as if they were alive." They traveled at speeds of about 1,000 miles per hour, could change their density at will, "like jellyfish," and moved erratically. "On their sides, upside down, right side up. Any human would be torn to pieces by the G's." He remarked that "nobody knows [if they're from another planet]...perhaps 'otherworldly' is the right word," and stressed that he was not the first to sight them, merely the first pilot to report one. "Ezekiel supposedly saw one."

The U.S. government has never publicly acknowledged the existence of UFOs, but reports of clandestine military investigations into the phenomenon continue to appear. Many pragmatic theories — such as the presence of illuminated swamp gas — have been put forward to explain them away. In 1951, the Office of Naval Research revealed that it had been using huge balloons for cosmic-ray research since 1947, the year of Arnold's sighting, and that the balloons could conceivably be mistaken for flying saucers. Not everyone accepts such prosaic explanations, of course. Speculation, sightings, and reports of contact of all kinds continue to proliferate. A clearinghouse called UFO Contact Center International, based in Federal Way, Washington, claims that 33% of Americans have had contact with aliens, although the center says only 1% has acknowledged such contact publicly.

Aerocar

1949

Sky Commuter

1985

The perfect vehicle, in many minds, would be one that could fly in the air or travel on land with equal ease. Remember the magic car from Ian Fleming's children's book, *Chitty-Chitty-Bang-Bang*? Ideally, a pilot in a flying car could, when encountering bad weather, simply land, remove the craft's wings and tail, and drive the vehicle through the storm.

Since the turn of the century, when both the automobile and the airplane were in their infancies, a number of pilots and engineers have tried to construct such a hybrid craft. The earliest recorded attempt was by airplane pioneer Glen Curtiss, who, in about 1917, crossed a Canard triplane with a custom-built "car"; the result was drastically underpowered, far too heavy to fly, and was soon abandoned. The Zuck Planemobile, built before World War II, was a single-seated, three-wheeled machine with a 40-horsepower Continental engine and a single pivoting wing; it proved too complex and expensive to build for mass production. Other prewar experiments, similarly unfruitful, included Bill Stout's Skycar, which had a rear-mounted "pusher" propeller, and Waldo Waterman's Arrowmobile, which combined a Studebaker engine and a flying-wing, tailless body.

The most practical of the early "roadable planes" was Robert E. Fulton Jr.'s Airphibian, which was certified by the Civil Air Authority in the late 1940s. It was a four-wheeler, powered by two electric aircraft batteries; when its wings and tail were detached, it became an automobile capable of 40 miles per hour. But the wing-tail assembly was so fragile that it had to be stored in a hangar and could not be towed behind the car, which meant that any flight had to return to the airport of origin in order to dismount the wings. In the minds of the public and potential investors this defeated the main purpose of a flying car, and Fulton's vehicle never went into production.

Several Northwest inventors have dabbled in "air cars" over the years. The dean of the fraternity is Moulton B. ("Molt") Taylor of Longview, who has devoted most of his life to perfecting the Aerocar, "the car with the built-in freeway." It was directly inspired by Fulton's Airphibian, but had a tail and wings that could be detached and towed behind the car. Although Taylor received certification in 1956 from the Civil Aeronautics Administration, he has yet to find financial backers to help him produce the Aerocar.

Born in Portland in 1912 and raised in Longview, Taylor became intrigued at an early age by itinerant barnstorming aviators, and took his first airplane ride at the age of 14. After becoming a private pilot and earning engineering and business degrees at the University of Washington, Taylor entered the Navy as a flight cadet and during World War II worked on missile technology. On his own time he designed an amphibious air drone, and, following his release from duty in 1946, he formed a company to produce a sport plane, the Duckling. While negotiating in Delaware for factory space, however, Taylor met Robert E. Fulton Jr., saw Fulton's Airphibian — and glimpsed his own future.

Taylor moved back to Longview in 1948 and, with his father's help, found 50 investors willing to put up seed money to design and build a prototype flying car. Within

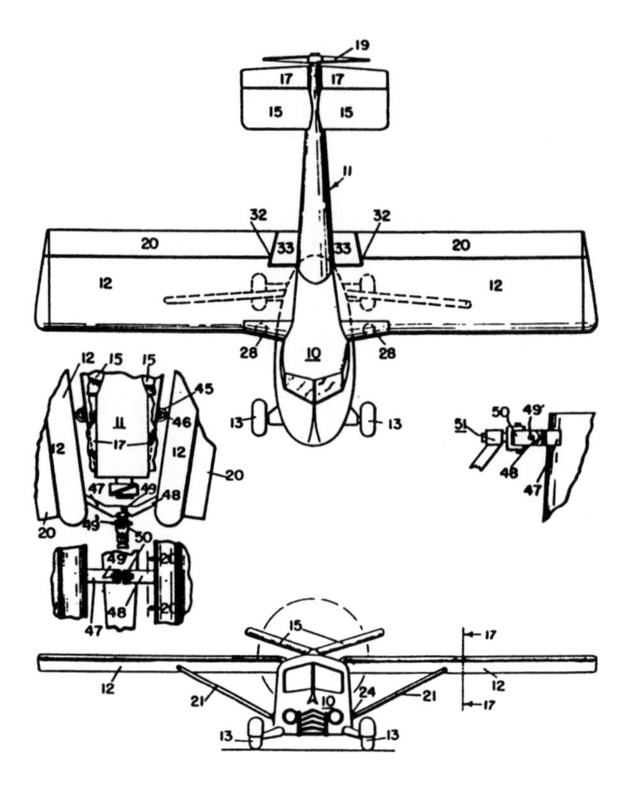

Molt Taylor's Aerocar followed in the slipstreams of previous "flying cars" — such commercial failures as Glen Curtiss's Autoplane, the Zuck Planemobile, Bill Stout's Skycar, Waldo Waterman's Arrowmobile, and Robert E. Fulton Jr.'s Airphibian. The Aerocar's detachable wings and prop were designed to let the pilot soar high above pesky traffic jams or drop down and drive through nasty weather.

nine months, aided by engineers Charlie Kitchell and Art Robinson and mechanic Jesse Minnick, Taylor had a prototype that successfully completed a test run from Longview to nearby Kelso, traveling one way by land and returning by air.

Two popular vehicles of the time, the Ercoupe two-seater airplane and the Crosley automobile, were Taylor's design paradigms. To make the automobile portion of their vehicle look as "normal" as possible, and to improve visibility and weight distribution, the team incorporated a high wing, a single engine, a three-speed manual clutch, and four tires. A "pusher" propeller, rear-mounted at the end of a long, upswept tail cone, eliminated the need to remove the prop before driving. Taylor also gave the craft an extra-heavy-duty suspension system, four-wheel brakes, controls and lighting for both air and land, an adequate heater, roll-down windows, windshield wipers, and air and land radios. Front-wheel drive kept the rear wheels, which touch down first in a normal landing, free from "spinning up" and damaging the differential and transmission.

The biggest single design problem Taylor faced was in dampening the torsional resonance of the long drive shaft from the engine to the prop. The power pulses of a reciprocating engine can cause tiny torsional twists in even the stiffest driveshaft, and when these get out of phase with other vibrations in a vehicle the shaft can break. After several unsuccessful tries, Taylor discovered a French invention called Flexidyne, a "dry fluid" coupling. It used tiny steel shot enclosed in a housing and packed to a nearly solid mass by centrifugal force. The shot provided just enough "give" to absorb the engine's power pulses.

Taylor received national publicity for his flying car even before it was certified, and he appeared in many national auto and air shows, magazines, and newspapers. Mobil Oil sent him on a promotional tour, he appeared on several TV news programs, and he was a guest on the popular TV game show "I've Got a Secret." (His secret was, of course, "I drove here in a car that flies!") During the next several years, Taylor raised $750,000 — just enough to build five prototypes, including one that was destroyed as part of the testing procedure. The Aerocar was certified by the CAA in 1956. Certification took a long time because so much new ground was being covered, and because Taylor was operating on a small budget. In those days the CAA was in many ways more cooperative with and sympathetic toward new ventures than its present-day counterpart, the FAA; it is unlikely that today a shoestring operation like Taylor's could reach the certification stage.

The four prototypes of the first Aerocar sold for $25,000 apiece, not a competitive price and much more than Taylor's initially envisioned figure of less than $10,000. (In contrast, a new 1956 Bonanza four-seater airplane sold for about $22,650.) Over the years, these four models were each sold several times over; in fact, Taylor partially supported himself by buying back airplanes (he had a clause in the sales contract giving him first right), then reconditioning them and selling them again. One highly publicized sale was to Los Angeles actor-comedian Bob Cummings, who bought an Aerocar in 1960 and featured it on his weekly TV show, "The Robert Cummings Show." Cummings, who had flown since the 1930s, had several airplanes; they were all named *Spinach* and were painted various shades of green. Cummings' Aerocar was bright yellow and green, the colors of Nutra-Bio, a vitamin company in which the actor, a health-food advocate, had an interest.

By the 1960s, both the car and airplane sections of the Aerocar had become outmoded, and Taylor revamped it to be comparable with the Cessna 172 and the then-new Ford Pinto. This design, the Aerocar III, included a more modern body style, torsion bar suspension to replace the old coil-spring suspension, and improved landing gear. In the 1970s, the Ford Motor Company expressed some interest in producing it, but the oil crises of the decade put an end to their speculation. The Aerocar III is now on display at the Museum of Flight in Seattle. It has a top speed in the air of 135 miles per hour, a top road speed of 60 miles per hour, a gross weight of 2,100 pounds, and a 500-mile cruising range when equipped with its standard 32-gallon fuel tanks.

Although slowed by a recent stroke, Taylor is still raising production money and working on a kit version of his latest model, the Aerocar IV, which is designed to attach directly to a standard automobile, specifically the Honda CRX. He expects it to sell for $150,000, not including the car. The Honda was chosen for its gas mileage, size, and for design factors such as its front-wheel drive. Taylor hopes the use of a standard automobile with road-tested and certified engine, brakes, and other parts will eliminate many of the problems he faced with previous Aerocar models.

Taylor's latest flying-car design requires only minimal modification of a standard Honda CRX.

A more recent Northwest entrant in the sky-car field is Fred Barker, whose company, Flight Innovations, is based at

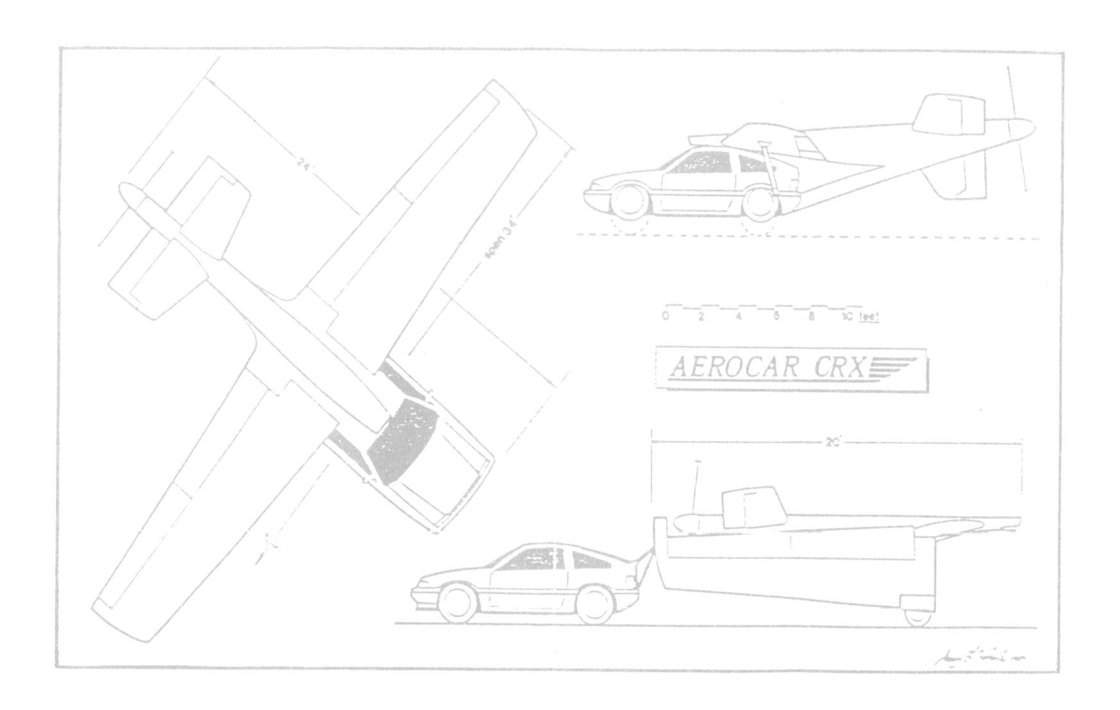

Arlington Airport in Snohomish County, Washington. Barker and his colleagues, including his son and daughter, are building the prototype of a sleek, high-tech, two-passenger vertical takeoff and landing (VTOL) craft called the Sky Commuter. The 8' × 14' Commuter, which features a hummingbird logo on its tail, should be capable of hovering in place, cruising slowly, or soaring at high speed. Power will come from four variable-speed electric rotors: three for upward lift, one for thrust. Its estimated top speed will be 85 miles per hour, with a top altitude range of 500 feet. Unlike conventional aircraft, it will not stall at lower speeds; instead, it will be capable of cruising as slowly as five miles per hour. (As of this writing, the Commuter has risen only about eight feet in field tests.)

Barker, a graduate of the University of Washington engineering school, has worked for Boeing and an oil company in the Middle East. In Saudi Arabia in 1984–85, he indulged his childhood fascination with dragonflies and other hovering winged animals ("anything that took off straight up, basically") by building a small radio-controlled model of a VTOL craft that simulated a hummingbird's flight. This became the inspiration for the first Sky Commuter, which Barker built in 1985 and 1986 after his return to America. His company currently operates a machine shop that produces precision parts for Boeing and other clients, which helps maintain cash flow during the long and expensive process of developing the Sky Commuter. Barker estimates he will need to build 14 prototype vehicles to get certification, and will spend $25 million before production begins. So far, the company has 80 stockholders who have invested in Barker's plan.

The name "Sky Commuter" is something of a misnomer. Although Barker does hope eventually to sell his invention

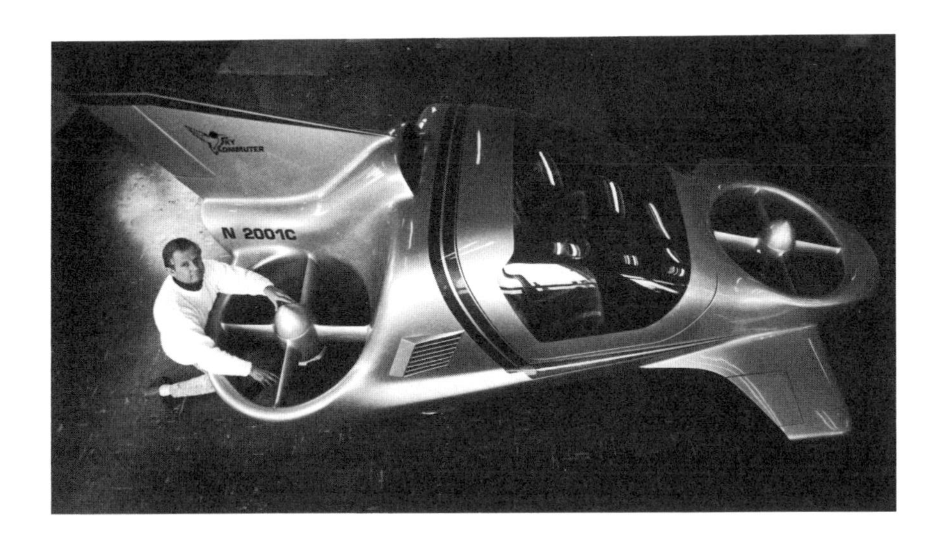

The Sky Commuter, a jet-age entrant in the flying-car sweepstakes, will have a Kevlar body, four 140-horsepower electric motors, and a computerized "fly-by-wire" guidance system.

to private parties as a true commuting vehicle, his immediate sales targets are military organizations and drug interdiction/border patrol groups who could use a fast-moving aircraft that hovers like a helicopter, speeds like an airplane, handles like a race car, and looks like George Jetson's Ferrari. When Barker does move into the private market, he hopes to sell the Sky Commuter as a home-built kit. It would be the most sophisticated kit-built aircraft available anywhere — as well as the most expensive, with an estimated price tag of $100,000 exclusive of engines.

Monorails

1910/1912

Beltway

1962

Seattle's Monorail had some imaginative, if mostly imaginary, predecessors.

In 1910, a Seattle inventor and entrepreneur named W. H. Shepard suggested an elevated monorail system as an auxiliary to offset the problems of streets clogged with horse-drawn and automobile traffic. Many other inventors were working on similar designs at the same time; what set Shepard's design apart was its provision for the carriage's center of gravity below the single rail; the system depended on guide rails on either side to steady it. Passengers would enter from raised platforms high above street traffic, and power would come from gas engines or electric motors; each car would be 30 feet long and 10 feet wide, with seating or standing room for about 40 people per car. Since the rails were to be made of wood, rather than metal, the system's cost was estimated at $3,000 per mile, considerably less expensive than the heavy rail or other forms of light rail proposed at the time. *The Seattle Times* commented in 1911 that "the time may come when these wooden monorail lines, like high fences, will go straggling across country, carrying their burden of cars that will develop a speed of about 20 miles an hour." Shepard and a few backers founded the International Monorail Company, set up a machine shop on First Avenue South, and produced a portion of track that was tested along the Seattle tideflats; the financial backing to complete the project fell through, however, and the plan fell into obscurity.

This 1911 photo shows a system typical of the schemes proposed in the early days of Northwest mass transportation. Note the open windows!

At roughly the same time, another Seattleite, John Wheeler, was spending the fortune he had made in the Alaskan Gold Rush in an effort, begun in 1912, to promote a monorail system that would run between Seattle and Olympia. (Earlier, in 1908, he had founded the Nooksack Railroad Company, which had tried but failed to build a railroad between Vancouver, B.C., and Bellingham.) A section of Wheeler's monorail was built and tested on the Tacoma mudflats, but Wheeler, like Shepard, was never able to interest enough backers or government officials in his project to make it viable. Even after his money was long gone, Wheeler remained a true believer, fighting to promote his idea until his death in 1933.

The Monorail, built for the 1962 World's Fair in Seattle, runs only from Seattle Center to downtown — hardly the grand scheme envisioned by the proponents of the Metro-Belt.

In 1962, just as the Seattle's World's Fair and its Monorail were attracting attention, came another mass-transit proposal, this one for a series of endless beltways similar to the moving sidewalks found in some airports. Engineer J. E. Schrock and accountant R. D. McPhaden formed a company, Metro-Belt Inc., to promote their idea of a series of "horizontal escalators" around the city. The system, which they claimed would be capable of accommodating up to 52,800 passengers at a time, was to be suspended on a cushion of compressed air, driven by pneumatics, and enclosed in tubes 20 feet above the ground. Several belts, ranging in speed from 8 to 40 miles per hour, would lie side by side and would be available to travellers moving in either clockwise or counter-clockwise direction; these passengers could move from a "slow lane" to an "express lane" by simply stepping from one belt to the next. (The fastest belt would be equipped with an endless bench.) Walls would separate belts going in opposite directions, with safety guaranteed by continuous closed-circuit television monitoring.

Riblet Tramway

1896

The Riblet Aerial Tramway was an early system of elevated cable cars invented by Byron C. Riblet of Spokane. His firm, Riblet Tramway, became one of the most prominent tramway companies in the world. Frank Bartel, in the *Spokane Business Journal* of February 28, 1966, wrote: "From the frozen reaches of the Yukon to the steaming jungles of India and the dizzying heights of the Himalayas, there are people in many lands whose lives are linked to Spokane by strands of steel."

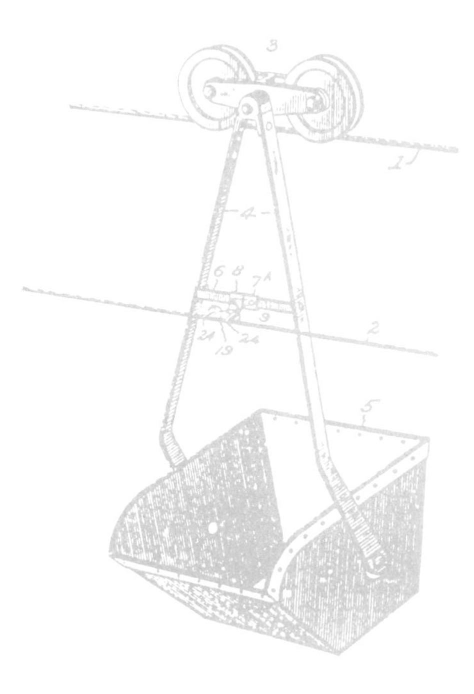

This 1906 "Tramway Bucket and Rope Clip" is just one of Byron Riblet's many successful designs for the mining industry.

Riblet, born in Iowa, graduated in civil engineering from the University of Minnesota and worked for the Northern Pacific Railway before settling in Spokane. There he formed his own general engineering business, specializing in electric railways and dams. His first aerial tramway was produced for the Noble Five Mining Company near Sandon, B.C., in 1896, and after its success he concentrated on the building of tramways. (In the 1930s, the company also branched out into the making of ski lifts.) The Riblet Tramway Company was incorporated in 1911, and the many patented refinements and improvements Riblet produced to augment existing tramway designs include a 1906 chair to carry miners to high mine-entrances. (Riblet had noticed that miners rode inside the ore buckets on his tramways anyway, so his invention simply made their lives a little safer.)

Riblet took in his younger brother Royal as part owner of the business, but in 1933 the two had a falling-out and Royal was ousted. He started his own tramway company, but it was not successful. For many years Royal and his wife

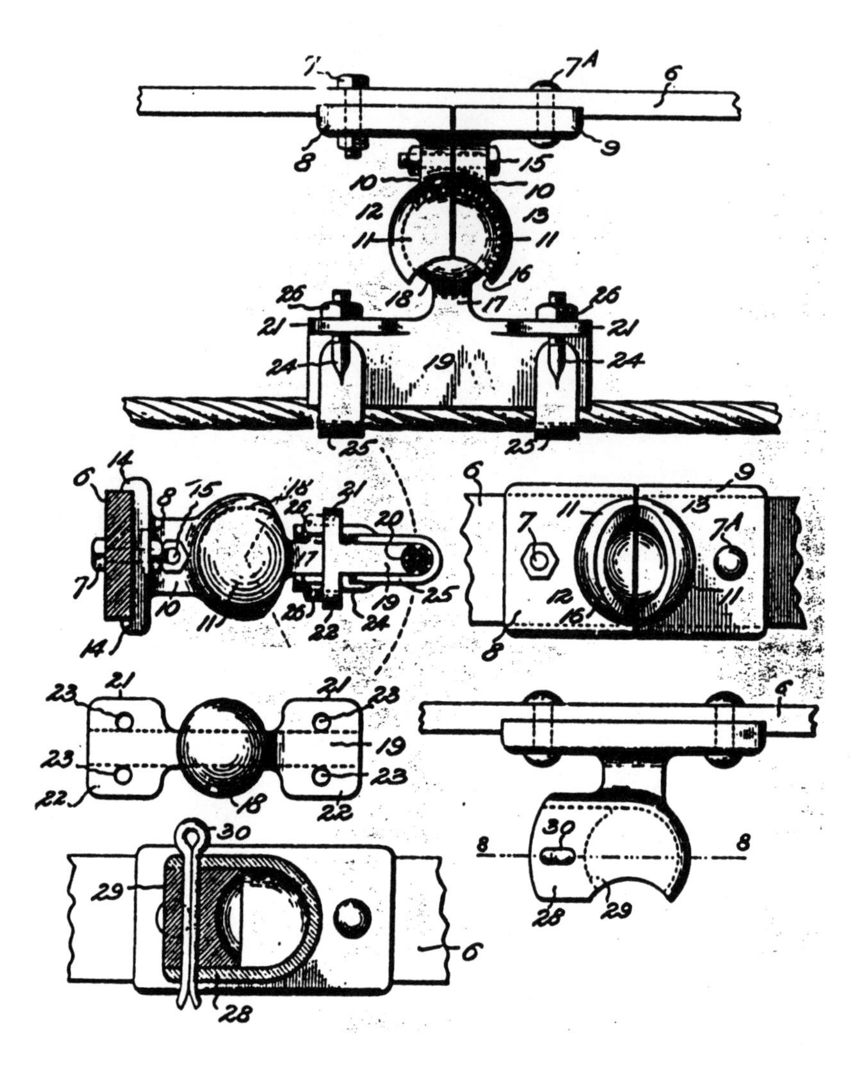

A detail from the patent drawings for Riblet's tramway bucket.

lived in a prominent "castle" on a bluff overlooking the Spokane Valley, equipped with a private 1,900-foot tramway that was a Spokane landmark until it was removed in 1956. Although there is some confusion as to who invented what within the Riblet company, it is known that among Royal's inventions was a 1933 design (never marketed) for a mechanical parking garage, consisting of a "live chain" of parking stalls that hauled cars up to different levels of a building along the outside wall.

High-Tech Wizardry

Simplified Typewriter Keyboard

1936

Perhaps the most striking example of a worthy Northwest-born idea that never received its due is the simplified type-writer keyboard invented in 1936 by Dr. August Dvorak of Seattle. Dvorak's system, a more efficient and less fatiguing alternative to the keyboard that has been the world standard for over 100 years, received considerable attention when it was introduced, but then languished, nearly forgotten, until the advent of computers and computer-related injuries led to a renewed interest in it. If Dvorak's design had met with initial commercial success, the typewriter keyboard we now know so well might now seem as quaint as the ETOAIN SHRDLU layout of an antique Linotype machine.

The QWERTY keyboard, named for the first letters in its top row, was designed by Christopher Latham Sholes, who created the first practical typewriter in the 1870s. "Practical" is a relative term; Sholes deliberately designed it to be as slow, inefficient, and cumbersome as possible, because the mechanism of his early typewriters simply could not keep up with the speeds some typists could reach. The keys would jam and break if worked too hard. This, coupled with Sholes's feeling that his "type-writing" machine should maintain the leisurely pace of handwriting, about 20 words per minute, led him to devise a layout that would deliberately hamper a typist's efficiency. He chose QWERTY precisely because of its many design faults, such as the fact that the weakest fingers of the left hand (the weakest hand for most typists) are responsible for many important moves, and that frequent

If the gods of typing had smiled on Dr. Dvorak, everyone today might be pounding out 150 words a minute to his simplified tune. This scene shows a typing class at Seattle's University Book Store in the early 1940s.

moves off the "home row" (the line where the fingers normally rest) are inevitable. QWERTY was adopted by Remington and other early manufacturers, and it quickly became the accepted standard.

August Dvorak, born and educated in Minnesota, was a professor of educational psychology at the University of Washington when, in 1932, he first became interested in improving Sholes's clumsy affair. (Although a distant cousin of the composer Antonin Dvorak, the professor pronounced his last name "Dvor-ahk.") Over the next several years, Dvorak devised a new keyboard system by studying the physiology of the hand and analyzing the most common English words and "digraphs," sequences of two letters which form written or spoken sounds, such as "th" or "sl." He was aided in his research by two grants from the Carnegie Foundation, by the pioneering ergonomic studies of Frank and Lillian Gilbreth (about whom the comic memoir *Cheaper by the Dozen* was written), and by his brother-in-law, Dr. William Dealey, an expert typist.

Dvorak called his invention the Simplified Keyboard. It later became known as the Dvorak Keyboard or American Simplified Keyboard, and today, in a modified form, it is sometimes referred to as a "language-based" keyboard. It was designed so that over 3,000 words, about 70% of the most commonly used words in English, can be found on the "home row," where they are most easily reached — as opposed to fewer than 100 words found on the QWERTY home row. Dvorak's system also divides strokes more evenly between the hands than QWERTY and takes into account common digraphs, strategically placing keys so that the combinations can be swiftly and easily executed.

The results are dramatically better than the QWERTY norm. The Dvorak keyboard is less fatiguing; tests conducted by the Navy in the 1940s show that during an average eight-hour day, a 40-word-per-minute typist's hands travel about 16 miles on a QWERTY keyboard while, with the same text, a Dvorak typist's hands travel about one mile. Dvorak allows greater speed; 90% of the world's typing speed record-holders use the Dvorak keyboard, regularly hitting speeds of over 180 words per minute with complete accuracy. (World champion Barbara Blackburn clocks in at over 200 words per minute.) Dvorak advocates point out further that the system is easy to learn; sixth-graders type 40 words per minute after

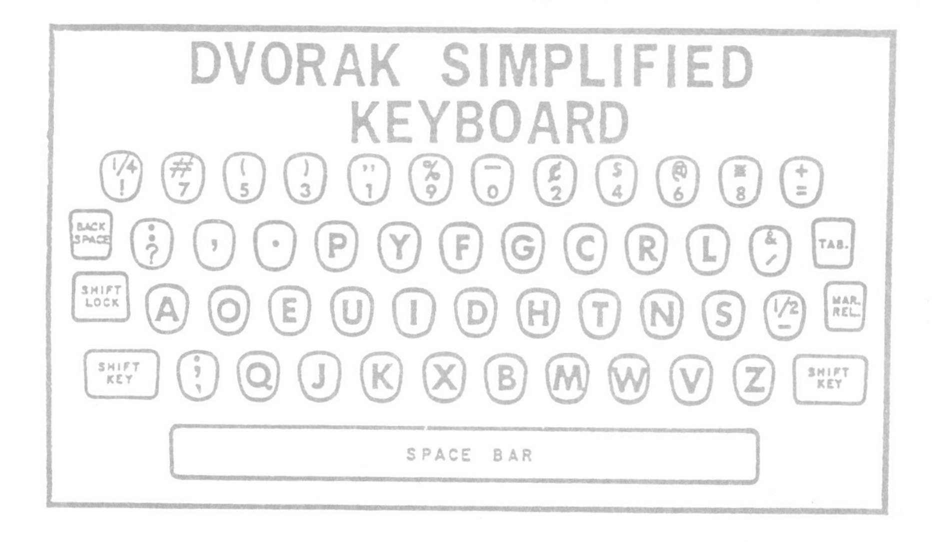

The simplified keyboard was designed to give our hands a fair shake.

only a few weeks of training. Adults average about 18 hours to reach 40 words per minute with Dvorak; QWERTY requires an average of 56 hours to do the same.

Dvorak's new typewriter created quite a stir when it was introduced. Typists set 27 world records for speed and accuracy within nine years, and contemporary newspaper articles expressed high hopes for the worldwide adoption of the professor's clearly superior method. "With such a flash [of publicity], I thought surely the public would accept it,"

Dvorak told a *Seattle Post-Intelligencer* reporter in 1974.

But it was not to be. Typewriter manufacturers and dealers, wary of the new machine, mounted a strong campaign against it. They feared that sales of a more efficient machine would mean fewer typewriters sold and (so the argument ran) more people out of work. Why should offices use two machines and two typists when they could do the same work with only one of each? They also cited the costs of modifying existing machines, and of retraining people to use the new system, even though Dvorak had designed his keyboard so that existing machines could be easily converted. Dvorak's own finances also conspired against him; America was still deep in the Great Depression, and after making a series of errors in business judgment, the professor lacked the capital to fight the status quo.

The Dvorak system went the way of Esperanto. Periodic efforts to revive it, such as the now-defunct Dvorak International Federation or the modified keyboard manufactured by Smith-Corona in the mid-1970s, failed to turn the public's head. But the serious repetitive-strain injuries (RSI), such as Carpal-Tunnel Syndrome, that are often suffered by keyboardists have created a renewed interest in Dvorak's invention. RSI may well become the number-one occupational hazard of the 1990s, and Dvorak advocates argue that Dvorak's greater efficiency could drastically reduce the incidence of RSI. One estimate of typists in the United States now using the Dvorak keyboard puts the number at about 10,000. In 1982, the simplified keyboard was accepted by the American National Standards Institute as an "alternate standard" to QWERTY, an important step in its legitimization. Some large organizations are experimenting with it, including the state governments of New Jersey and Oregon and various telephone directory-assistance centers. Computer technology, which can handle the speed of a human typist, is also part of the reason for the Dvorak system's comeback. Software programs to modify standard keyboards and specialized classes are making Dvorak easier to learn. Some computer hardware and software packages, such as Microsoft's Windows 5.0 package and the Apple IIc computer, come with built-in Dvorak capability.

Several high-tech, non-Dvorak keyboards have recently been created by Northwest researchers specifically to address the problems of RSI. Among these are the Kinesis keyboard, invented by Seattle medical-technology designers William Hargreaves, Shirley Lunde, and William Ferrand, and the Maltron ergonomic keyboard, manufactured in England but coinvented by Seattle designer Jack Litewka. Both these keyboards have the traditional QWERTY layout but use such ergonomically sophisticated features as palm supports, separated and specially angled left/right sections, and even optional audio tones that let operators know when it is time to take a break.

Color Television Tube

1944

While teaching physics at the University of Southern California during World War II, Dr. Willard Geer became seriously ill. His weakened condition and mounting medical bills also brought on a severe depression. When his wife half-jokingly said that he should invent something to make them rich, thinking that it might help him overcome his depression, Geer took her at her word. He explored several possibilities for "really big inventions" and eventually settled on one: color TV.

Experiments with color TV had been conducted as early as the mid-1930s, and NBC had begun the first regularly scheduled black-and-white broadcast in 1939, but in the mid-1940s television was still a largely experimental medium, even in black-and-white. (The first regular color broadcast did not occur until 1953.) Although she had no previous training in science, Mary Wells Geer began doing research for her husband, abstracting articles in the UCLA library during the day while he taught, so that he could study them in the evenings.

Black-and-white television operates, very simply, through emissions from an electron gun. When these are picked up by phosphates on a monitor screen, the phosphates glow and an image is formed. Color TV requires three separate emissions, representative of the three primary colors, that must be blended in precise amounts to produce an image. How to blend them? Geer's breakthrough came in early 1944, as he was walked up a flight of stairs in the USC physics building. As he watched sunlight falling on both treads and risers, he realized that a serrated screen would provide the answer. Electrons beamed from different angles from three guns onto opposing serrations would hit only one side of each pyramid-shaped serration. If each side of each pyramid row were covered with a particular phosphor, the beam striking that phosphor would emit only that particular color. The result, seen from the front, would be a color picture. At the breakfast table the next morning, Geer demonstrated the idea to his wife and two sons by sticking rows of sugar cubes into the sides of a block of margarine, forming a series of pyramidal shapes above and below the margarine "screen." The Geers built a working prototype of a three-

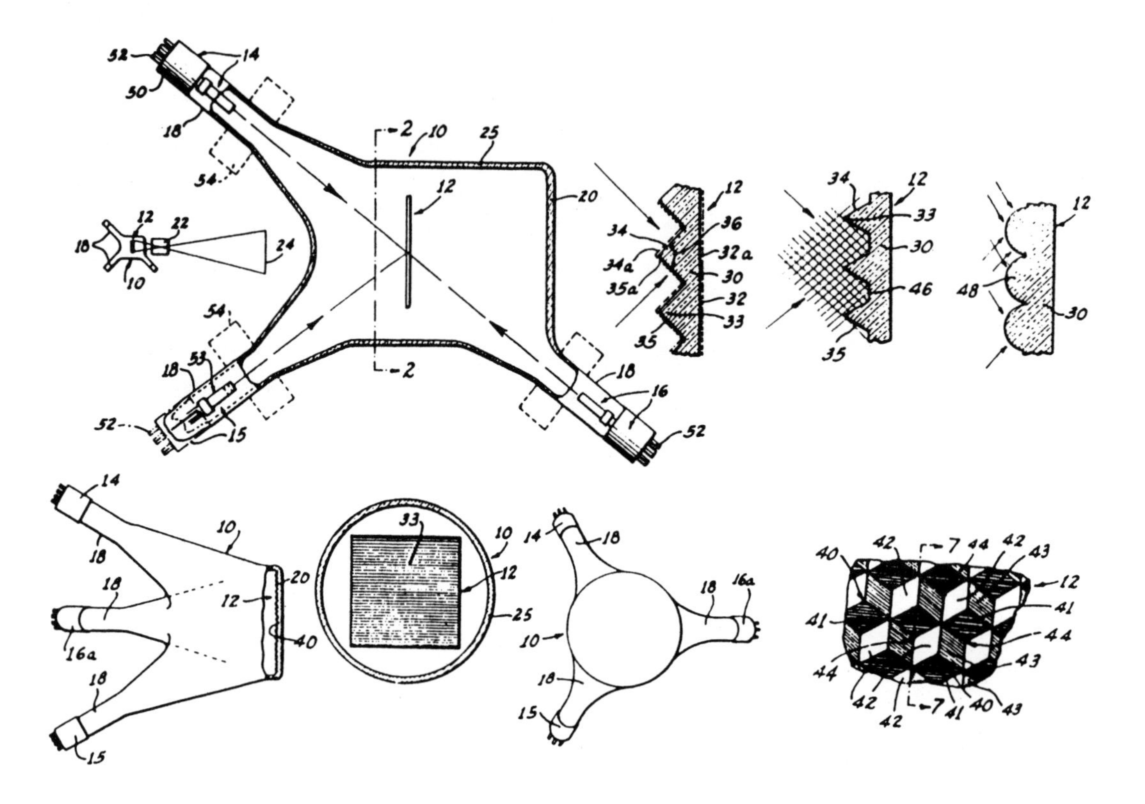

Only one of Willard Geer's color picture tubes was built, but its underlying principle is a major component of the system used today. A model of that prototype, built by the Stanford Research Institute for Technicolor, is on display in the Smithsonian's TV collection.

gun tube in their kitchen, using such scavenged tools as impression wax from the office of a sympathetic dentist, a sewing machine bought for five dollars from Goodwill, and paintbrushes to apply phosphors to the tiny pyramids by hand. In her unpublished memoir, *Why Not Invent Color TV in Your Kitchen?*, Mary Geer wrote: "Much of the conversation at family meals centered around what we could do about color television. Perhaps it had become a family obsession. By this time the pursuit had become as exciting as any wild game hunt, with a feeling that the animal was big." Geer found a patent attorney who was willing to handle his invention on a contingency basis, and the claim was filed on July 11, 1944. On August 5, the RCA Corporation filed a remarkably similar patent, then lodged an "interference suit" against Geer. Four years of bitter litigation followed. Geer was eventually vindicated, awarded his patent in 1949, and given $15,000 by RCA to settle. (Geer owed $10,000 to his attorney, and Geer's annual teaching salary at the time was $3,800. Mary Geer says, diplomatically, that the settlement was "decent, but not as much as it should have been.") RCA eventually made a deal with Technicolor to develop a version of the Geer tube, modified so that one gun emitted all three beams, the principle used in most color TVs today.

Although the Geers' pioneering work took place in California in the mid-1940s, they both grew up in the Northwest. They returned to the Seattle area in 1967. "Doc" Geer subsequently taught physics at Bellevue Community College, and that school's observatory was named in his honor following his death.

Oscilloscope Technology: Tektronix

1947

Tektronix, the world's leading manufacturer of oscilloscope equipment, has always been a major source of innovation in its field. "Tek," as it is known, was founded in 1946, and the two key players in its early years were Jack Murdock, an appliance salesman, and Howard Vollum, an engineer who had performed distinguished work with the Army Signal Corps in developing new techniques for radar observation. Both were Portland natives, just out of the service, and

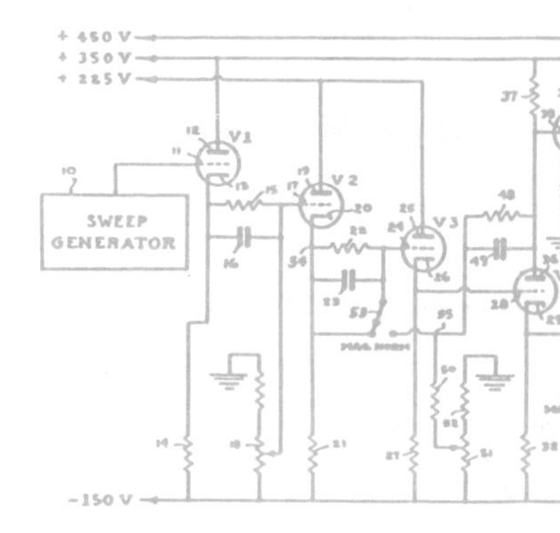

eager to capitalize on postwar surplus products and scientific innovations.

Oscilloscopes test a variety of electronic equipment for such problems as incorrect resistance and faulty wiring or components, by converting an electronic signal into a visible, wavelike electron beam projected onto a screen. The particular shape of a given wave indicates certain electric values, which can be measured against the screen's grid and adjusted by changing the visible wave shape. Oscilloscopes were not new when Tek was founded; the cathode-ray tube, the electron-shooting "gun" that is the machine's heart, was invented by Karl Ferdinand Braun of Germany in 1897, and the first practical "oscillograph" was invented by William Duddell of the United Kingdom in that same year. But even in the years immediately prior to World War II, oscilloscopes were produced by only a few companies, such as RCA and Dumont, and those that were available were crude, clumsy, oversized affairs, with extremely low sensitivity, accuracy, and versatility. Unless specially modified, prewar 'scopes were unable to record transient, random events (that is, one-time-only signals), which made them useless in many situations. What was needed was a device that could measure more than simply repetitive or continuous electronic events. Wartime needs had advanced oscilloscope technology to the point of including such innovations as a triggered sweep circuit, similar to that of radar's, and a built-in repetition generator that triggered both signal and observation mechanisms simultaneously. Armed with knowledge of these advances, Vollum became convinced that a better commercial oscilloscope could be produced than the crude mechanisms then on the market.

Two Portland entrepreneurs, Howard Vollum and Jack Murdock, launched the Tektronix Corporation by building a better oscilloscope with surplus World War II equipment.

With the help of Milt Bave, a mechanical engineer who joined Tek early on, Vollum created the 511, Tektronix's first oscilloscope, in 1946. Production began soon after. (Vollum's earlier prototype, the 501, was sophisticated but huge, weighing over 100 pounds and standing several feet high and wide; Bave helped him shrink it to a compact, manageable size and weight.) The 511 represented the first commercial use of a radar-style triggered sweep mechanism for an oscilloscope, and it was the first commercial machine to have a calibrated amplifier and a calibrated time base. It was also considerably more sensitive, accurate, and lightweight than any oscilloscope then on the market. Tek's first customer, in 1947, was the University of Oregon Medical School.

The company's first products were manufactured on a typical shoestring budget: the first 511s were made with surplus parts, and the partners borrowed the use of the *Oregonian* newspaper's sheet metal fabrication facilities to make the chassis and box. Assembly was done by the partners themselves, along with their wives and anyone else who happened to wander nearby. Despite this humble start, the company's product was so superior to anything else available that it became a quick success. As production increased, other models soon followed that incorporated further innovations, including ceramic-strip technology for easy assembly (an early example of modular construction, developed by engineer Ted Goodfellow in 1951). The company's first patent was for technology related to sweep magnification, and was filed in 1953 by engineer Dick Ropiequet. More recent innovations include a ceramic cathode-ray tube, a direct-view bistable storage tube (DVST) that allows images to remain on screen for long periods, and technology for extreme miniaturization of parts.

Sonic Holography

1971–

The world of high-end stereo equipment is volatile and fast-moving, with new products appearing and disappearing daily. Audiophiles — "hi-fi nuts" — are a fiercely opinionated and ultracritical breed. In this rarefied world, a Snohomish, Washington, man named Bob Carver has emerged as one of the industry's most controversial and enduring designers.

Carver is also prolific. In the October 1988 issue of *Ultra-High Fidelity*, Gordon Brockhouse states: "It's unlikely that any audio designer alive has more innovations to his credit than Bob Carver." His companies, Phase Linear and the Carver Corporation, have successfully marketed audio gear that is both affordable for the average listener and challenging to audiophiles who appreciate higher levels of performance. Although most of his innovations occurred before compact discs and digital technology eclipsed LP records for all but the most diehard purists, Carver's work continues to be an important influence.

Carver's first successful product, in 1971, was the lightweight and inexpensive Phase Linear 400, the best-selling high-power amplifier in the world until it was superseded by his Phase Linear 700. (The prototype for Carver's first amps was a 700-watt model he'd built in college, at a time when 150 watts was considered the upper limit. He housed the model in a large coffee can because he couldn't afford a fancy shell.) Other successful designs include a noise reduction system, the first of its kind, that worked for LP records as well as tape recordings.

But Carver is probably best known for a device called

the Magnetic Field Amp, and for hardware developed to create the phenomenon he calls "sonic holography." In 1977, Carver sold Phase Linear; the next year, he started the Carver Corporation. Since he needed two new inventions to make a splash on the market, he decided one would be an inexpensive, lightweight power amplifier and the other a way to create "acoustic holograms" — the illusion that sound is originating all around the listener. Mindful of the value of timely publicity, Carver introduced his ideas to the world before they were actually invented: "I went to New York before the big consumer electronics shows and invited writers to come talk with me. I was still in the failure mode on one of [my products], and for the other I hadn't the faintest notion what I was going to do. I patented them as soon as they were designed, of course, but at that point I hadn't designed them yet. I was bluffing — I went to the show with an empty box. But I knew I could do it." The bluff worked; Carver generated considerable publicity, delivered on his promises, and turned both items into smashing successes.

In the late 1970s, high-quality 100- or 200-watt power amps were expensive, with virtually nothing available under $1,000. They were also heavy, ranging anywhere from 35 to 55 pounds. The primary reason for the excessive weight was that these big amps required large power supplies and a lot of heat-sink metal to carry away the heat they generated. For over a year, Carver worked alone in his Snohomish home lab to develop a lightweight power source. He was unsuccessful until he hit on the idea of lowering the frequency, storing the energy in a magnetic coil, and using a switching device that worked like a light dimmer. By "reading" instantaneous musical demands and drawing only as much power as was needed at the moment, the amp's size could be reduced fivefold.

"When I realized that, I sat bolt upright in bed one night — it was literally like a flash out of the blue — and I roared off to work. After a year and a half of beating my brain and pounding my butt against something that wouldn't work, within hours I had a proof-of-concept model, and within seventy-two hours I'd gone from proof of concept to a working prototype." The resulting production model, the M-400 Magnetic Field Amp, weighed less than 10 pounds and retailed for $349. One *Stereo Review* critic called it so revolutionary that he predicted it would "have the same influence on future amplifier circuitry that the introduction of the acoustic suspension technique had on loudspeaker system design."

Around the same time, Carver was also working on his acoustic holography idea. The main reason that a listener can distinguish a conventional musical recording from the real thing is that psychoacoustic "cues," normally present with a live event, are missing on recordings — or, more precisely, there are too many of them. Every sound event in real life produces two signals, one in each ear, that arrive a few microseconds apart. The brain receives one signal slightly after the other, and this temporal displacement tells the brain where the sound is coming from. But in a stereo playback, that same sound will produce four such sonic arrivals, two per speaker per ear, and the brain knows that the sound is coming from two separated loudspeakers.

Carver compares the process to looking at a picture: "A photograph has all the colors and hues, but it doesn't have depth information unless you use a binocular viewer. Certainly, with photographs you can have a great deal of enjoyment — the colors, tones, textures, all the emotional things — but it's obviously different from looking out the window, because it doesn't have the depth. Music's the same way." His solution

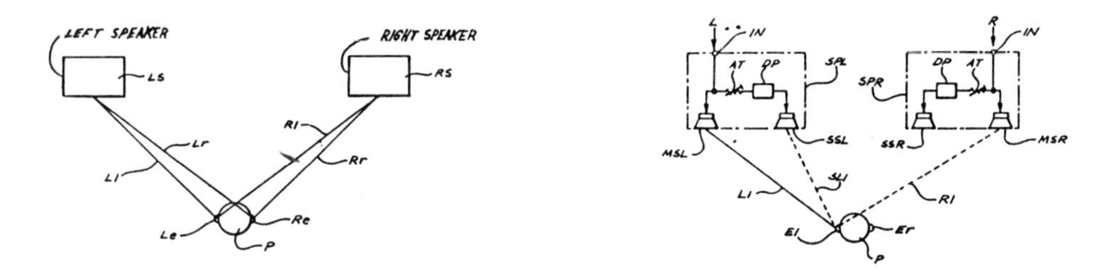

Bob Carver's innovative high-end audio equipment creates the illusion of 3-D sound.

was to supply the missing depth information by erasing the two extra signals with acoustic wavefront cancellation (sound patterns that intersect and erase other unwanted signals). The ear again hears only two cues and accurately reconstructs the sound field in the brain — somewhat like a visual hologram, in which two beams of light intersect and, by virtue of destructive and constructive interference patterns, form an apparent three-dimensional image.

Carver's current project ("*What* I'm doing is no secret; *how* I'm doing it, now that's a secret") is the next evolution of his acoustic hologram. "The problem with the hologram is that to make it work you have to sit in the perpendicular bisector of the loudspeakers before its magic happens. If you move your head off axis more than a half-meter or so on either side, the illusion degrades into plain stereo. You can always tell if people are listening to sonic holography — they're all lined up in a row. This new hologram uses an extra pair of speakers in the front that generate the hologram signal, so instead of creating the interference pattern in the air around your head, it creates it between the loudspeakers. The virtual three-dimensional images are therefore formed near the loudspeakers, instead of near your ears, so you can sort of walk around the images."

BASIC for the Microcomputer

1975–77

MS-DOS

1978–81

BASIC (an acronym for Beginner's All-purpose Symbolic Computer Instruction Code) was first developed at Dartmouth College in 1964. It was a vast improvement in speed and facility over FORTRAN, the comparable computer language then in use, but it was written for large, expensive, complex machines, in an era when computers were still exotic animals. Although the personal computer revolution was waiting to happen, before things could explode the movement needed a language that could run on a simple, practical, low-cost machine. BASIC for the microcomputer was that language and, in 1975, Bill Gates and Paul Allen created it. Their program became the industry standard and was a major player in the home-computer revolution. Gates and Allen used it to launch Microsoft, now the largest and most influential software company in the world, and later used their experience to help develop the world's most popular software program: MS-DOS, an operating system used with over 60 million computers.

The story of how Gates and Allen developed their program is the stuff of legend among hackers (computer experts who write their own programs). The two had been friends and hacking colleagues since their high school days in Seattle, and in 1974 were living in Boston, where Allen worked for Honeywell, and Gates, at 19, was a Harvard student. Late that year they saw the cover of the (predated) January 1975 *Popular Electronics*, which featured the Altair, a computer developed by Micro Instrumentation Telemetry Systems (MITS) of Albuquerque, New Mexico. The Altair was probably the first true "personal" computer, a crude machine by today's standards but revolutionary for its time. It had no keyboard, monitor, or permanent data storage — but it did have a tiny memory and a low price, $397. Gates and Allen decided to write a version of BASIC for it. In an interview in the book *Programmers at Work*, Gates recalls: "Paul Allen had brought me the magazine with the Altair on it, and we thought, 'Geez, we'd better get going, because we know these machines are going to be popular.' And that's when I stopped going to classes and we just worked around the clock."

They contacted Ed Roberts, the president of MITS, who told them that other hackers had similar ideas. Roberts said he'd buy from whoever demonstrated a usable model first. Allen and Gates worked on their program day and night for three and a half weeks, armed with a manual about the 8080 microchip, the Altair schematics, and a large computer that ran a program Allen had written to simulate the Altair. In their book *Fire in the Valley*, Paul Freiberger and Michael Swaine write: "There was no established industry standard for BASIC or for any other software. There was no industry. By deciding themselves what [their version of] BASIC required, Gates and Allen set a pattern for future software development that lasted about six years. Instead of researching the market, the programmers simply decided, at the outset, what to put in."

When Allen arrived at the MITS office in Albuquerque, he had a paper tape containing his and Gates's program. When he loaded it into the Teletype reader connected to the only machine at MITS that had the requisite 4K of memory, no one knew exactly what would happen. But the program worked. The Teletype read his information and replied, "MEMORY SIZE?" Allen pressed the "ENTER" key, typed in "4K," and then typed a request: "PRINT 2+2." The machine gave the correct answer — "4" — and Allen knew he had the beginnings of a successful program.

Loading from a paper tape was a primitive method of programming, however. In late 1975, MITS decided to produce a more sophisticated disk version, which would eventually become the world's first retail floppy-disk system. In early 1976, Allen, by then Director of Software for MITS, asked Gates to write it. According to legend, Gates, still a Harvard student, checked into the Albuquerque Hilton with a stack of yellow legal pads and came out five days later with a new disk-based version of BASIC. After five more days of work at MITS, he had it up and running.

In *Programmers at Work*, Gates remembers: "We just tuned the program very, very carefully.... The really great programs I've written have all been ones that I have thought about for a huge amount of time before I ever wrote them.... So by the time I sat down to do [what became] Microsoft BASIC in 1975, it wasn't a question of whether I could write the program, but rather a question of whether I could squeeze it into 4K [of memory] and make it super fast. I was on edge the whole time thinking, 'Will this thing be fast enough? Will somebody come along and do it faster?' "

In the embryonic days of computing, enthusiasts lived by a nebulous code called the Hacker's Ethic, which held them responsible for sharing new software by making free copies available to any interested party. It was felt that advances in software could only better the world and should therefore be common property. Allen, Gates, and MITS president Roberts felt differently. They believed that developing software represented as much work as developing hardware, and that programmers should be compensated accordingly. Allen and Gates formed a partnership called Microsoft and licensed the rights to their BASIC to MITS, which began charging for the program and paying Microsoft a royalty for each one sold.

Microsoft modified BASIC in 1977 into a version for NCR and in 1978 adapted it for CP/M, an 8-bit operating system developed by Digital Research of Pacific Grove, California. Over the winter of 1978–79, with their program on its way to becoming the industry standard, Gates and Allen moved their base of operations back to the Northwest and opened shop in Bellevue, east of Seattle. (Microsoft is now headquartered in nearby Redmond and Gates is still its head. Allen has formed his own Bellevue-based company, Asymetrix, which is marketing a project called Toolbook.)

Microsoft's success with BASIC continued with MS-DOS, which has become the standard operating system for the 16-bit computer technology that superseded the 8-bit systems. (MS-DOS is an acronym for Microsoft Disk Operating System. An operating system manages the flow of data between a computer and peripheral devices like terminals, printers, floppy diskettes, and hard disks. It also manages computer files, loads programs into the computer's memory, and initiates program execution.)

MS-DOS was written in 1978 by Tim Paterson, a programmer working for a company called Seattle Computer Products. That year, Intel Corporation released its 8086 microchip, significantly faster and more powerful than the 8080 used in the Altair. Paterson began work on a CPU (central processing unit) card that could run on it. Gates and Allen, meanwhile, had a hunch that 16-bit technology was the wave of the future and were working on the development of a version of BASIC for the 8086 chip. Paterson finished his

8086 card in early 1979, and by late spring had an operating CPU; working with Microsoft, he soon had their BASIC program running on it. This early 8086 operating system had several names for its various incarnations, including Q-DOS (for Quick and Dirty Operating System) and 86-DOS.

Until the late 1970s, IBM, the world's largest manufacturer of mainframe computers, had virtually ignored the microcomputer market. When the company decided to enter the personal computing market in 1980, it was interested in developing hardware around Microsoft products. Microsoft agreed to produce a series of languages (BASIC, FORTRAN, Pascal, and COBOL) for a still-prototypical 16-bit IBM computer, as well as an operating system. Microsoft had an operating system for BASIC but needed similar systems for the other languages, and Paterson's 86-DOS fit their requirements. Gates and Allen negotiated in the fall of 1980 with Paterson and Rod Brock, the president of Seattle Computer Systems, leasing it for $50,000 and a series of license-back considerations.

Over the winter, Paterson and Microsoft programmer Bob O'Rear adapted 86-DOS to the prototype IBM hardware. (IBM kept Paterson largely in the dark for security reasons while he was still employed by SCP; he was not allowed even to see the prototype machine until he became a Microsoft employee in May 1981.) Accounts vary as to the amount of day-to-day work Gates and Allen put into developing 86-DOS or its modifications, but it is clear that they were overseeing it carefully and, at the same time, preparing a masterful marketing campaign. The new system made its public debut in August as PC-DOS, version 1.0. Modified for PC-compatibles (computers manufactured by other companies but similar to IBM PCs), the program was given the name it retains today: MS-DOS.

Desktop Publishing

1984

"Desktop publishing" refers to the use of computer software programs to design and print a variety of graphic material. The procedure is widely employed, using relatively simple techniques and inexpensive software and personal computers, to create high-quality material such as newsletters, brochures, and catalogs. Prior to the mid-1980s, the process was known as "electronic publishing" and required expensive equipment and highly skilled operators. As a result of recent user-friendly programs, however, desktop publishing is rapidly becoming a standard office procedure.

The term "desktop publishing" was coined at the 1984 Seybold Conference on Electronic Publishing in San Francisco by Paul Brainerd, founder of the Aldus Corporation in Seattle. Brainerd had previously worked for ATEX, which manufactures computer hardware and software for large-scale print operations such as newspapers. He started Aldus in 1984 with the idea of adapting ATEX's electronic publishing techniques for microcomputers. His PageMaker program, first introduced in 1985, was not the first true desktop publisher; several others, including one called Quark Xpress, appeared on the market shortly before or nearly simultaneously with it. But PageMaker was the first really practical program of its type, and has remained the most popular.

Voice-Activated Computer Translator

1989

Inventor/entrepreneur Stephen Rondel of Redmond specializes in high-tech tools for frequent travelers. Among his products are a one-cup drip coffee maker, a portable smoke alarm, a travel steam iron, a portable safe, and a razor that removes "pills" from clothing. Rondel, who has a background in computers, recently turned his attention to what he thinks is the ultimate problem facing travelers: the need to break language barriers. His solution is VOICE, the world's first hand-held, voice-activated computer translator.

Introduced in 1989 and selected by *Popular Science* as one of that year's greatest achievements, VOICE recognizes and translates 35,000 words and phrases. It weighs about three pounds, measures eight by seven by three inches, and contains one 8-bit and two 16-bit microprocessors. There is no keyboard; the machine's exterior has only a single button, an LED display screen and a small speaker. If one needs to translate English into French, for instance, an English-French cassette is inserted into the machine and the user dons an earpiece with a small microphone (VOICE has difficulty distinguishing ambient sound from the user's voice, so the microphone must be held near the mouth). When an English phrase is spoken, the LED screen displays it in English so the user can make sure the computer has received it correctly. Then the user presses a button and the computer repeats the phrase in French. (Its synthesized accent is clear but seems vaguely Scandinavian no matter what language is spoken.)

The last word in high-tech tools for travelers: a voice-activated, hand-held computer that translates words from one language into another.

Rondel came to voice-activated translation in a roundabout way. Equipped with a degree in systems engineering and several years of experience as a computer expert, teacher, and lecturer, his original plan was to enter the computer field directly. He got sidetracked into the world of high-tech travel tools, however, when in the late 1970s he acquired the rights to and redesigned a power supply that adapts electrical appliances to overseas voltages. After the 1980 MGM Grand Hotel fire in Las Vegas, a client suggested that he make a portable smoke alarm. Rondel developed the item, but dropped it because he couldn't produce it cheaply enough for the average consumer. (Another Rondel-patented item, a portable door lock, has also not gone into production because of the expense.)

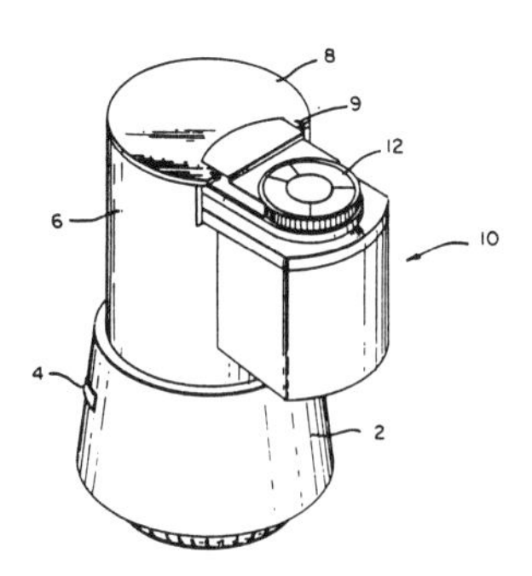

The one-cup drip coffee maker was the first patent Rondel received, in 1985. "Coffee makers are very large consumer items. People drink a lot of coffee, and there was no such thing as a traveling coffee maker. I usually start out [inventing] with an absurdity; in this case, I felt that any responsible cup should be able to make its own coffee. The ironic thing is that I stopped drinking coffee about the time our first maker came off the line."

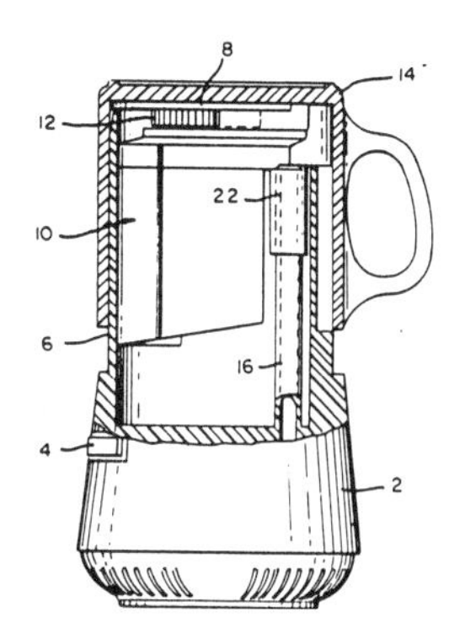

Rondel's most financially successful item so far has been the Garment Groomer (also known as Lift-Off), a product he "hardly took seriously" at first. He had been developing a travel shaver when an affiliated company sent him a shaver with extra-large holes in the head and told him to try it on his clothes. "I gave it to my director's wife, and she tried it on her cashmere sweater. I thought for sure it was going to eat the sweater, but she loved it — and then we started taking it quite seriously. It didn't require a lot of effort [to redesign and market], but it did very well."

In 1985 Rondel felt the time was right to develop a

hand-held computer translator. Computer-synthesized speech had been available for several years, but voice recognition — in which a computer recognizes and responds to spoken commands — was still a cutting-edge concept, and a practical, consumer-oriented machine was, at the time, only a distant possibility. His first prototype had a keyboard, which Rondel decided was superfluous. "I think it's beneath the dignity of a human being to speak to a computer with your fingertips. To have all the resources of a computer, and then to have to dedicate all your time and your own most valuable resources — your hands and your eyes — to making it respond to you, is a giant waste. I'd much rather talk to a computer, and have it talk back to me, as easily and as fluidly as two people talk."

When Rondel first began work on VOICE, many experts thought that miniaturizing a translation computer was impossible within our lifetime. Rondel says, "Now, we can give [skeptics] the benefit of the doubt; they might have meant that a machine couldn't be built that would translate every word I could say. But there is a market for something that does less than the whole English vocabulary, and we showed it was possible to build a machine that would do just that." For Rondel, the possibilities of "conversational computing" extend far beyond translation. He would like to create a VOICE application that lets an engineer "talk through" a problem-solving session with a computer while leaving the engineer's hands free to work. ("I have this thing in my hands and it'll rotate only thirty degrees or so." "Try doing this...") Another possibility would be a voice-activated word-processing program that would let writers "talk" their text directly into a computer without the need of a keyboard. Rondel is currently applying for a patent on the system that will operate such a program.

Northwest Fun and Games

The Electric Guitar

1931

Several pioneers in the use of amplified instruments claim the invention of the electric guitar as their own. The history is murky at best, and the question of who *really* invented the one instrument that best symbolizes rock and roll may never be settled. Among its early pioneers were John, Rudy, and Emil Dopera, three Czech immigrant brothers who formed the National Guitar Company in Los Angeles in 1926 to manufacture a steel-bodied guitar that was much louder and brighter-sounding than wooden-bodied instruments. (The brothers split from National two years later and formed another company, Dobro, which manufactured a variation on the National theme: wooden-bodied guitars with built-in metal resonators, known as Dobros.)

There were others. Les Paul, who later became famous through his recordings with Mary Ford, began designing solid-body amplified guitars in the late 1930s. Clarence ("Leo") Fender, a radio repairman and amplifier maker who branched out into making steel guitars and pickups and brought electric guitar–making to a pinnacle with his famous Stratocaster and Telecaster models, received the first patent for an amplified guitar, in 1944. Adolph Rickenbacher, who owned the tool and die company that made steel bodies for National, also experimented with an early electric guitar and formed a company called Electro-String (later changed to the simplified spelling of Rickenbacker).

These early Tutmarc electric guitars were a revolution waiting to happen.

There is a reasonably good case, however, for citing a north Seattle basement as the birthplace of the electric guitar.

Paul Tutmarc was a Seattle music teacher, bandleader, singer, and Hawaiian-style guitarist, known on the Pantages and Orpheum vaudeville circuits as "The Silver-Toned Tenor." He loved to tinker with new varieties of guitars, and was especially interested in finding a way to increase the instrument's volume. A common complaint among guitarists during the big-band years of the 1920s was that their instruments were consistently drowned out by the other players. (The best solution yet proposed was the National steel-body guitar, which used its resonating metal diaphragm to amplify the sound.) In 1931, at the suggestion of a friend from Spokane named Art Stimpson, Tutmarc began experimenting with ways to electrically amplify guitars using a simple magnetic pickup, similar to the type of acoustic amplifier used in telephones. He made a large pickup from a horseshoe-shaped magnet wrapped with thin wire (probably number 38 or 40) and covered it with friction tape. This was placed inside an old round-hole, flat-top guitar and connected to a crude amp that Tutmarc had built from a radio with the help of an electronics repairman named Bob Wisner.

Whang!

Tutmarc had been expecting increased volume, but not the new dimension in sound the instrument produced. Not only was the guitar louder, it had a tonal range — bright to murky, chimelike to ragged and distorted — that had been impossible to produce or even conceive of with acoustic guitars. He quickly went on to amplify other instruments, including upright basses, pianos, and zithers. Further experiments with amplified guitars resulted in an octagonal solid-body guitar and a solid-body bass made of white pine, which were probably the first of their kinds, and a "cluster guitar" with seven necks tuned seven different ways.

Tutmarc never copyrighted his guitar. The Patent Office

Paul Tutmarc, "The Silver-Toned Tenor," once serenaded Jean Harlow with a rendition of "There's Danger in Your Eyes, Cherie." In this 1972 photo (left), he plays a Hawaiian-style guitar of his own design. Below: It took two strong men to lift one of Tutmarc's early amplifiers.

deemed it inadmissible because Alexander Graham Bell had long before patented the telephone pickup. "I blew about $300 in the patent offices," Tutmarc recalled in a *Seattle Times* profile shortly before his death in 1972. "Anyway... [the electric guitar] wasn't an invention so much as bringing existing forces together." Tutmarc lacked further money to pursue the patent question — a familiar story with Depression-era inventors — and simply began making and selling his amplified instruments without copyright protection, using the brand name Audiovox. He hired an employee of the Anderson and Thompson ski company, Emerald Baunsgard, to produce the fine woodworking and inlay work for his black walnut guitars on a part-time basis. When Tutmarc's son, Paul Jr., known as Bud, reached adolescence, he too was put to work, winding coils for pickups and assembling amplifiers. (Tutmarc's friend Stimpson, meanwhile, went to Los Angeles and sold the idea of the amplified guitar to the Dobro company for $600.)

Tutmarc's least expensive guitar sold for $39.50, including the case and a three-tube amplifier; his regular line went for $112.50, with a case and five-tube amp included. His bass amps, which used a 12-inch Jensen speaker instead of the 10-inch Lansing speaker used for the guitar amps, sold for about $110, while the basses themselves went for about $75 apiece. Advertisements for "Tutmarc custom electric guitars" began appearing in local newspapers, along with more prosaic announcements of Tutmarc's availability as a guitar teacher. Tutmarc sold his guitars as fast as he could make them, but it never became more than a sideline to his teaching and performing. Although no household-name musicians used Tutmarc's guitars, they proved popular with traveling show bands and church groups, particularly the lightweight electric bass he'd invented because he "felt sorry for the bass player, who always had to take his big bass along in the car and was forced to drive by himself to the next show."

The 1937 yearbook for John Marshall Junior High, Bud Tutmarc's alma mater, contains an ad for a Tutmarc electric bass, which Tutmarc Jr. played in the school band. (Bud has the distinction of being the first high-school kid in the history of the universe to own an electric guitar. His father reflected in the 1972 *Times* profile: "A lot of fathers and mothers probably would like to kill me. Then again, if it hadn't been me, it would have been somebody else.") Bud went on to build his own line of amps, bass guitars, and guitars. Today he is a real estate broker in Seattle and the owner-operator of a small recording studio. He and his son also carry on the family tradition of performing Hawaiian-style music for audiences around the world. Bud still plays the last guitar his father made, finished sometime in the 1950s; he estimates that an original Tutmarc would sell today for at least $2,500, but adds that "there isn't enough money in the world to take [my father's guitars] from me."

Thirty-five years after Paul Tutmarc began his experiments, another Seattleite, Jimi Hendrix, added his own set of inventions to the canon of the electric guitar.

More Musical Notes

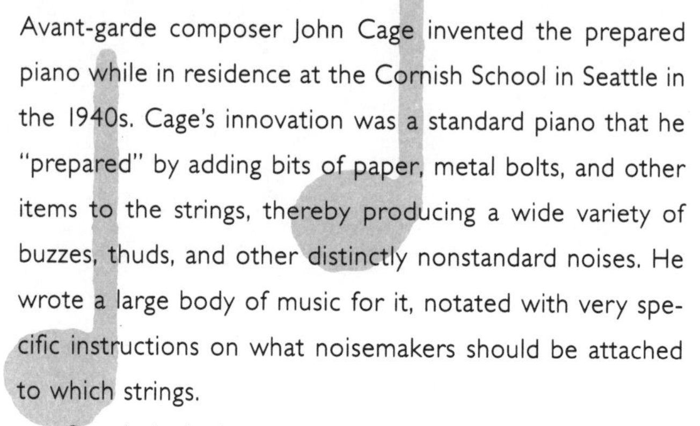

Avant-garde composer John Cage invented the prepared piano while in residence at the Cornish School in Seattle in the 1940s. Cage's innovation was a standard piano that he "prepared" by adding bits of paper, metal bolts, and other items to the strings, thereby producing a wide variety of buzzes, thuds, and other distinctly nonstandard noises. He wrote a large body of music for it, notated with very specific instructions on what noisemakers should be attached to which strings.

Seattle is, by happenstance, the elevator music capital of the world; the three largest "business music" companies, including the Muzak Corporation, are all headquartered there. Chief among the innovations in this field that have come out of Seattle is the concept of "foreground music" — that is, the use in a background-music setting of original tunes by original artists, as opposed to the syrupy versions usually associated with "audible wallpaper." The foreground idea was pioneered and successfully marketed in the early 1970s by Seattle native Mark Torrance. His company, YESCO, spawned two other major firms (AEI and EMS) before it merged with the Muzak Corporation. Muzak, AEI, and EMS are all still based in Seattle.

Otto Lagervall of Yakima invented the Manhasset Music Stand, the number-one-selling music stand in the world, in the late 1930s while teaching music in the Manhasset, Long Island, school system. His stand differed from previous models primarily in that it used friction-held clips to maintain proper shaft height, rather than the clumsy thumb-screws used previously. Lagervall patented the stand in 1940, and the device, which he manufactured in his basement, became popular around Long Island and beyond. In 1942, he returned to his childhood home of Yakima, set up a small factory in a garage, and continued making music stands until he sold the business to a friend in 1959. In his later years, Lagervall became intensely interested in kayaking, traveling the world and, in the process, inventing about a dozen kayak-related accessories.

One Northwest instrument that never went very far was the product of three anonymous Seattle musicians. The trio is credited in a 1936 *Popular Science* article with the invention of a cylindrical "merry-go-round" harp called a rondolin. Its strings were arranged along a five-foot-high vertical spindle powered by foot pedals, enabling a four-octave range to shoot past a performer's fingers, all within easy reach. Its inventors predicted great things for it, envisioning a day when the rondolin would be standard equipment for all orchestras. It was not to be.

GOLFING IN SPOKANE: 1960s/1976

In the 1960s, Dr Ernest G. Burnett, a Spokane optometrist, devised a series of weights for golf clubs that enabled a golfer to practice with a heavier club than normal to develop muscles; they also made the golfer's regular clubs feel light by comparison. In 1976, two other Spokane residents, Steve Gaffaney and James Henderson, patented a "robot golf teacher" that resembled a soft-drink vending machine but was in fact a hybrid instant-photo machine. It took stop-action pictures of a golfer's swing; by analyzing the photos, a golfer could get a better idea of how to improve strokes.

Neither the weighted club set nor the golf "teacher" was a financial success.

Erector Set and Chemistry Kits

1909

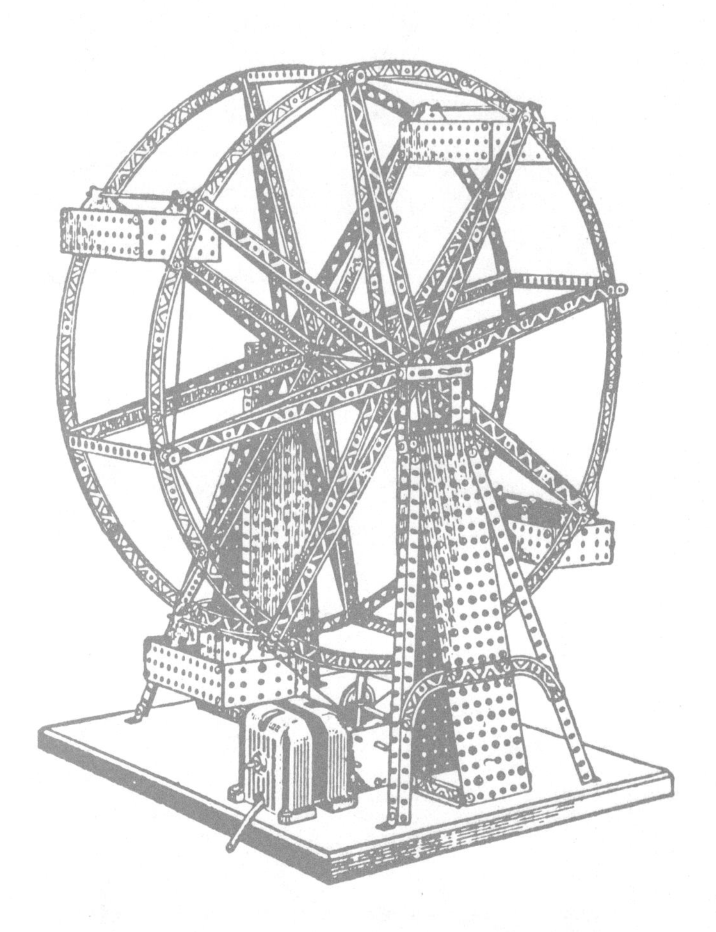

Hello, boys! Make lots of toys!

For generations, children have spent countless rainy afternoons building cranes, trucks, Ferris wheels, and other elaborate toys from the component parts of the Erector Set. The Erector Set and the many other toys manufactured by A. C. Gilbert's Mysto Manufacturing Company, including the American Flyer train and various magic, chemistry, and telegraph kits, combined fun with the wonders of science. They typified the pro-science attitude toward children's playthings that prevailed in America at the turn of the century and afterwards, in contrast to the simple wood building blocks and hobby horses of previous times.

As a college student, A. C. Gilbert set a world's record for the pole vault and participated in the 1908 Olympics.

Alfred Carlton Gilbert was born in Salem, Oregon, and early on showed a fascination with magic, machinery, and gadgets. As a teen, Gilbert was an athlete, setting a world's record for chin-ups while at preparatory school in 1901 and another for the running long jump the following year at Oregon Agricultural College (now OSU). He supported himself while studying medicine at Yale by performing magic shows, and made his first boxed sets of magic tricks, which he sold for five dollars each, after unsuccessfully trying to teach magic to friends who were unwilling to spend hours practicing. (While at Yale, Gilbert also set a world's record for the pole vault and participated in that sport during the 1908 Olympic Games in London. He never finished his medical studies.)

In 1909 Gilbert and a friend, John Petrie, founded the Mysto Manufacturing Company (later the A. C. Gilbert Company) in New Haven, Connecticut. The company sold an expanded version of Gilbert's early magic kits; these included coin and handkerchief tricks, a decanter that changed milk into coffee or water into ink, and tools for producing rabbits, flowers, flags, and coins from pockets and sleeves. In 1911, while riding the train from New Haven to New York, Gilbert noticed the Hartford Railroad raising new power lines in anticipation of the system's conversion from steam to electricity, and the towers inspired him to create the Erector Set. As he wrote in his autobiography, "I saw steel girder after steel girder being erected.... I found it interesting to watch their progress from week to week...it seems to be the most natural thing in the world that I should think about how fascinated boys might be in building things out of girders."

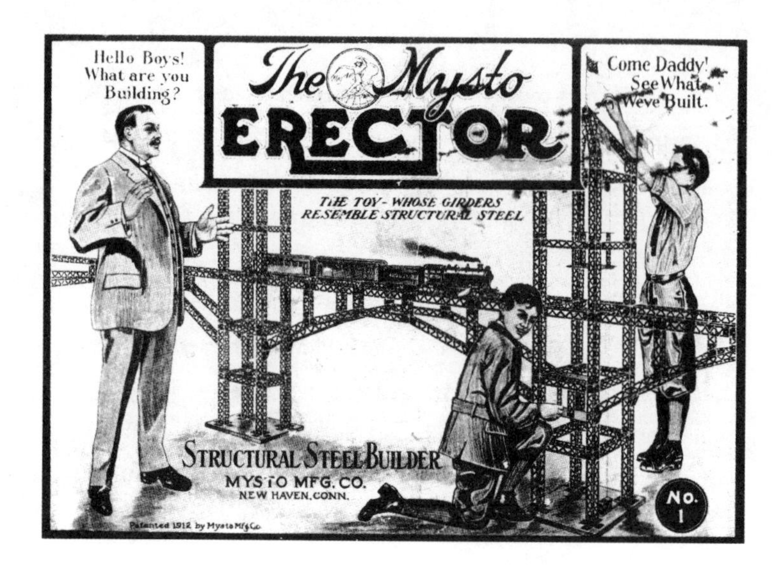

The inspiration for Gilbert's Erector Set came from his observation of railroad power lines in Connecticut.

Gilbert and his wife, Mary, began experimenting with cardboard girders, working with them until they had created a set of component parts that could fit together in many different configurations to form a variety of objects. A machinist then made Gilbert a prototype set out of steel, with locking edges and nut-and-bolt fastenings.

The Erector Set was introduced to the public in 1913 at the New York and Chicago toy fairs; two similar toy kits, the Meccano and the Richter Anchor Block, already existed, but neither had the complex locking edges, gears, wheels, pinions, or motors featured in Gilbert's set. Depending on their complexity, Erector Sets sold for between one dollar and $25; the most complex could be used to build windmills, Ferris wheels, drawbridges, and other intricate structures. The toy quickly became a success; over the next 20 years, 30 million Erector Sets were sold. Gilbert was the first toy manufacturer to advertise nationally, taking out full-page ads in magazines such as *Good Housekeeping* and the *Saturday Evening Post* as early as 1913. He also advertised his toys through a newsletter, "Erector Tips," which featured promotions such as contests in which boys won "diplomas" and prizes from the Gilbert Institute of Science and Engineering. The Mysto Company's aggressive ad campaigns included the familiar slogan "Hello, Boys! Make Lots of Toys!"

The Erector Set was Gilbert's most famous product, but he took out patents on 150 other inventions, many of them based on the small electric motor he had developed for the Erector Set. In 1916, he introduced the Polar Cub Fan, the first low-priced small fan in America; its small size was made possible by the then-new practice of coating electrical wires with enamel rather than wrapping them in bulky cloth. Other Gilbert appliances included the first electric coffee maker, the first hand vacuum cleaner, and the Kitchenetta, an early food processor that could beat, chop, extract juice, sharpen knives, sift, crush ice, and grind coffee. Educational toys continued to be a major aspect of Gilbert's work; one of the most popular was his chemistry set, introduced in 1917, which included a booklet called "Fun with Chemistry" and which let children create disappearing ink, matches with colored smoke, stink bombs, crystals, and glow-in-the-dark writing. (A Yale chemistry professor once surveyed his students and found that most of them had used Gilbert sets as children.) Other Gilbert toys included a microscope set, an intercom-telephone kit, a telegraph set, and a weather station.

Another toy that captured the popular imagination was the American Flyer Train. Gilbert bought the American Flyer company from a manufacturer named W. O. Coleman in 1939 and substantially altered the toy so that every piece was true to scale, modeled after real trains, and much more realistic. Gilbert invented a two-track system that eliminated the center rail, as well as the 3⁄16" S gauge, a more appropriate size for children than the HO (1⁄8") or O (1⁄4") gauge. William Smith, Gilbert's chief engineer, devised many of the

The Erector Set was usually a toy (here Gilbert shows off a familiar configuration), but occasionally it had more serious applications. Sir Donald Bailey used one to design his famous World War II–era Bailey Bridge, and the first successful heart-lung machine employed the kit's tiny electric motor.

Flyer's most famous features, including the mechanism that made smoke stream from the smokestack and the "talking" station that produced sound effects.

Gilbert produced one major failure. In the late 1950s he introduced the Atomic Energy set, which was unofficially encouraged by the U.S. government in the hopes that it would help the public understand much-feared nuclear power. It consisted of a Geiger counter, uranium ore, a cloud chamber where the paths of alpha particles could be observed, an electroscope to measure radioactivity, and a spinthariscope, an instrument that revealed radioactive disintegration on a fluorescent screen. The set was discontinued shortly after its introduction when parents began protesting about the danger, although a safer form of uranium was later incorporated into Gilbert's chemistry kit.

A museum in Salem, the Gilbert House Children's Museum, has memorabilia and examples of Gilbert's toys on public display.

Balsa-Wood Airplanes

1930s

N. E. ("Jimmy") Walker of Portland founded, in the early 1930s, American Junior Aircraft, also known as the A-1 Aircraft Company. Walker created many of the varieties of balsa-wood airplanes familiar to schoolchildren and model airplane fans in the years following. His designs and inventions ranged from five-cent gliders and rubber-band–powered planes to large models with gas engines, the semiautomatic "pistol plane launcher," and the "U-Control" line operator. (An example of his line-controlled Fire Bee model is on display at the Oregon Historical Society Museum.)

As a young man, Walker tried his hand at building full-size personal airplanes and worked as a machinist for Boeing in Seattle, then went home to Portland and turned to designing and manufacturing model planes. For many years he was the world champion radio-controlled–airplane operator.

American Junior Aircraft, the largest model airplane company in the world, is now headquartered in Portland.

Slinky Pull-Toy

1952

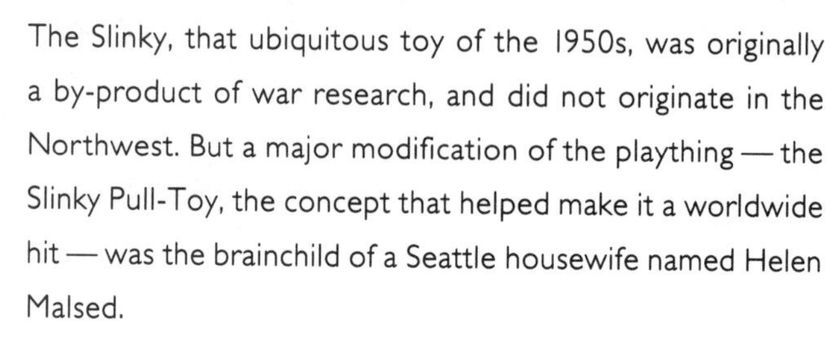

The Slinky, that ubiquitous toy of the 1950s, was originally a by-product of war research, and did not originate in the Northwest. But a major modification of the plaything — the Slinky Pull-Toy, the concept that helped make it a worldwide hit — was the brainchild of a Seattle housewife named Helen Malsed.

During World War II, Richard James, a marine engineer at the Cramp Shipyards in Philadelphia, searched for a way to keep heavy shipboard instruments, such as radios, from rocking too much during high seas. He collected hundreds of types of industrial springs and tried each as a ceiling hanger for the instruments. One day in 1943, when James accidentally knocked a particularly pliant steel spring off a pile of books, he discovered that it flipped over in a graceful arc, then flipped again, end over end. The spring was not much good as a shock absorber for nautical instruments, but James and his family began playing with the spring at home and imagining how it could be marketed as a toy. Toward the end of the war, he "borrowed" the design, and James Industries, based in Germantown, Pennsylvania, began marketing the "Slinky Spring." All it could do was turn end-over-end down a flight of stairs, but that was enough to make it a moderately strong seller.

A birthday present at the Malsed house set the Slinky pull-toy to wiggling. Left to right: Bob Porter, Rick Malsed, Bobbie Allen, Charlie Foss.

In 1952, Helen Malsed's son Rick got a Slinky Spring as a present for his sixth birthday. Rick Malsed recalls, "I looked at the toy and said, 'Boy, would that ever go if it had wheels!' The next thing I knew, my Dad was down in the basement burning himself with a soldering iron to try and put wheels on the thing. Then my mother attached a string she'd gotten from a window shade, and there it was — they'd made a prototype." With the addition of wheels, the Slinky took on a new dimension. No longer confined to simply arching end-over-end down stairs, now it could be pulled along the ground in a wonderfully wiggly, and quite unpredictable, fashion. Rick and his friends loved it, and his mother soon thought of dozens of variations that could be created by adding appropriate heads and tails to the wheeled spring — a caterpillar, a train, a dog, a hippo, a fire engine, and more. Helen Malsed wrote a letter to James Industries suggesting a licensing agreement, but "they were very conservative," she says. "They only wanted to use one of my ideas, and that was for the train."

Public response to the new train was moderate at first, but things changed when entertainer Arthur Godfrey singled out the toy for praise during his 1954 Christmas television broadcast. During the show, which was an annual staged party for his staff, Godfrey pulled the plaything from underneath a Christmas tree and charged around the studio, saying, "Isn't this just the cutest thing you ever saw!" After that, Richard James's factory couldn't keep up with the demand for the toy; in the next year, he increased his production eightfold by adding several of Malsed's variations to the original train idea. The Slinky Pull-Toy has since outsold every other kind of pull-toy worldwide.

Helen Malsed was not a novice inventor when she encountered the Slinky; although it was her most successful

Helen Malsed with some of her inventions — strictly for fun and games.

idea, it was neither the first nor the last. Even as a college student she showed an innovative streak: in the early 1930s, while at Whitman College, she started one of the nation's first newspaper shopping-chitchat columns, for the *Spokane Spokesman-Review*. It appeared every Thursday and she was paid five dollars per column. Her first invention came in 1942, when she was the assistant sales promotion manager for Seattle's Frederick & Nelson department store. At that time, it was the custom for sorority girls to pass around a five-pound box of chocolates at the dinner table to announce their engagement. The box would go around the table and when it had made a complete pass the girl would announce the groom's name. Malsed devised a custom-made chocolate box, with places for signatures on the lid, that could be kept as a souvenir of the occasion. She extended the idea to include special cigar boxes for new husbands and sold the idea for $500 to Gil Ridean, then manager of the bakery, candy shop, and tearooms at Frederick's.

She recalls, "Soon after that, a friend of mine told me I'd never get rich from cigar boxes. She was right, but what to do? We decided that since there was a baby boom coming, either clothes or toys would make sense. Well, I couldn't sew a stitch, so it had to be toys." Over 25 toys, some successful and some unsuccessful, followed: the "Toy Toter," a canvas holdall that, when unsnapped, formed a map on which children could play with trucks and cars; a board game called "Roman X," which Malsed believes is the only board game to use Roman numerals; and a number of pull-toys with moving parts, licensed to manufacturers such as Fisher-Price and featuring names like "Splashin' Myrtle Turtle," "Capt'n Tippy Tug," "Cat and Canary," "Pop-Up Bakery Truck," and "Klickety Klack," the first pull-toy to use clockwork.

Still a big item in today's baby stores is a toy that Rick Malsed believes is the only one of his mother's creations besides the Slinky Pull-Toy that he had any role in inspiring: "When I was in junior high, every girl had pop beads—small, brightly colored plastic beads that snapped together to form a bracelet or necklace. My mom saw these on some girls who were my friends, and she got the idea to make them into toys for babies, blowing them up so big that they couldn't be swallowed by an infant. She licensed the idea to Fisher-Price, and they still sell well."

The Happy Face

1967

Love it or loathe it, it can't be avoided: on lapel pins and wall posters, sanitary wrappers and RV tire covers, key chains and refrigerator magnets, the Happy Face is everywhere. Steven Heller, a New York–based graphic design historian, calls it "a genuine pop icon," one of the modern world's most recognized emblems — right up there with Kilroy, the peace symbol, and the red circle-slash that says no in any language.

Drawing a round face with two eyes and a smile is nothing new, of course; but the Happy Face *per se* is a relatively recent development. Its true origins will probably never be known (it has been ascribed to a New York novelty-item specialist, a Northeast insurance company, and a New York City radio station), but one of the most credible stories is that it was invented in 1967 by a Seattle public relations man as part of a campaign for a small bank.

The Vietnam era was a difficult time for the merchants of Seattle's University District. The street culture of University Way, with its mix of hippies, litter, revolution, and drugs, was driving most customers away. One organization in particular, University Federal Savings & Loan (now University Savings Bank), was especially sensitive to the tense climate; its office had been bombed once already. To lift community spirits and woo customers, the bank asked David Stern, a young advertising man with his own agency, to create a campaign with a double mission: to give the depressed neighborhood a boost, and to demonstrate that University Federal was a concerned community participant. Stern was mulling over a number of possibilities when one night he went to the movies, saw Dick Van Dyke in *Bye Bye, Birdie*, and walked out humming one of the film's most memorable songs, "Put On a Happy Face."

Stern, who is now semiretired after a career in advertising and an unsuccessful mayoral campaign, recalls: "It just stuck in my head. I drew this little character and found a theme: 'Open a savings account and put on a happy face.' I had my art director clean the thing up, we made one little button, and I put it on. That was the test-marketing — me, wearing it around. How many people would see it and smile? The answer was, a lot. So I ordered 100,000 of them, and the bank put them out in bowls." The reaction was swift. They were grabbed by the handful, and soon lapels all over Seattle sported beaming yellow faces.

The first hint that Stern's creation might take on larger proportions came months later, when an insurance company — then headquartered across the street from the bank — used the symbol in its annual report. Stern says, "I was not very happy, but we were stuck. The bank did business with the insurance company, and we couldn't upset them. I did tell their president that I thought they'd done wrong. His reaction, to quote another song of the day, was, 'Qué será, será.' I realized then that it was open season."

Stern never bothered to copyright his design. "In advertising you always come up with new ideas. You never know what's going to work, you're too busy to protect everything, and — unless you're terribly vain — you can't believe that everything you do will cover the world. If I'd tried to copyright it, I'd have spent the rest of my life in litigation. My kids love to remind me that if I had just two cents for every time the thing was used I'd be bigger than Donald Trump. But so what? I wouldn't have had time to watch those kids grow up."

Icosahedral Juggling Balls and Electronic Juggling Instruments

1977

The Flying Karamazov Brothers, a globe-trotting quartet of entertainers based in Port Townsend, Washington, specialize in what they call "juggling and cheap theatrics." The Karamazovs are not generally known as inventors, except perhaps as the creators of spontaneous wisecracks during their Marx-Brothers-on-steroids stage shows. Nonetheless, they can claim two important advancements in the tools of the juggling trade.

Although one of the Karamazovs' main claims to fame is the juggling of unusual objects (including "live" chain saws, frozen alligator heads, and chocolate cream pies), they often use traditional juggling clubs and balls. It seems logical, therefore, that the Brothers' first innovation was an improved juggling ball. In 1977, the group lived in a San Francisco apartment building that was also the home of Ann Worth, a seamstress and amateur mathematician. Worth's dual skills proved useful when Tim Furst (Fyodor Karamazov, the Brother

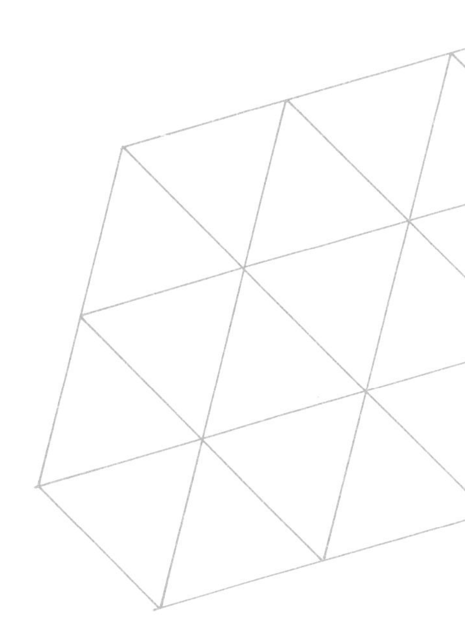

Chocolate cream pie, anyone? Lit torches? How about a Platonic juggling ball? Opposite, left to right: Tim Furst, Paul Magid, Howard Patterson, Sam Williams. This page: Twenty of these triangles are used in Ann Worth's pattern for the icosahedral juggling ball.

who remains silent on stage) asked her to help him develop an improved juggling ball. The existing implements were essentially modified beanbags, and Furst wanted something more spherical that was also easy and inexpensive to construct. Worth and Furst looked for inspiration at a variety of simple solids and at the topsy-turvy drawings of M. C. Escher. After some trial and error, with the other Karamazovs pitching in by subjecting prototypes to "torture tests," they settled on an icosahedron — one of the Platonic solids, a spherical shape made of 20 triangular panels. Other near-spherical juggling balls existed, but they were generally made with either baseball-style curved panels or with segments like a sliced orange. Both of these shapes would have required sewing difficult curves; the icosahedron, on the other hand, has only straight seams, except for two turns on one side, making it much easier to produce in quantity. Since cloth has a crisscross grain, and stretches if pulled diagonally, Worth designed a pattern that ensured that most of the ball's 60 seams were off the grain of the cloth, so that when it was sewn together and filled it formed a nearly spherical shape even though each panel was a flat surface.

Worth says, "I liked it because it was an elegant solution to the problem. It rounds out by itself, instead of us having to force it to round out." (Sam Williams, the red-haired Brother known onstage as Smerdyakov, adds, "Besides, Buckminster Fuller has nice things to say about the icosahedron.") Further experimentation proved that wild birdseed provided the optimum weight and shape for the stuffing. A variety of tightly woven cloths, mostly velveteen and bright cottons,

were chosen to cover the balls, and a company called Zen Products was formed to manufacture them. (The firm's name derives from an imaginary corporation that ran "ads" during the Karamazovs' early stage shows, extolling the virtues of such products as Consciousness-Raising Bran, Traditional Sex Rolls, and Mental Floss.) Although the design was never patented, it became quite successful; Worth estimates that since 1977 she has produced close to 100,000 juggling balls.

The second innovation attributed to the Karamazovs is a collection of devices known as electronic juggling instruments. These are a still-evolving set of tools, collectively developed in their Port Townsend rehearsal barn, that allow the Karamazovs to play live music by bouncing juggling clubs and balls off an array of electronic drum pads mounted on stands or on the jugglers' own bodies. Each time an object hits a drum pad, it produces an electronic impulse that travels through a battery of synthesizing and sampling equipment to produce a single musical tone, which in turn is broadcast over a loudspeaker system.

An early version of the electronic juggling arrangement involved drum pads attached to football helmets and chest protectors worn by all four Karamazovs; during a complex juggling routine, each Brother would strike a part of his body with a club at a precise moment, creating a single note each time, and the combined result became Beethoven's "Ode to Joy." The instruments have since become considerably more sophisticated: the devil-red jumpsuit currently worn by Howard (Ivan) Patterson during one segment of the show, for instance, has a dozen drum pickups sewn into various sections over the elbows, knees, and other sections of his body. Patterson is thus able to create his own complex, real-time musical compositions by juggling balls and bouncing them off his body.

Pickle-Ball

1965

Pickle-Ball has evolved from a slapdash summer entertainment for a few restless kids to an internationally recognized sport.

The game of Pickle-Ball was invented in 1965 by three friends, Barney McCallum, Bill Bell, and Joel Pritchard, who hoped to ease the boredom their kids sometimes experienced during summer vacations on Bainbridge Island. (According to legend, the game was named after the Pritchard family's cocker spaniel, Pickles; but Pritchard, then a U.S. congressman and now Washington's lieutenant governor, says he simply felt the new game needed a crazy name — the dog's name came later.)

McCallum, Bell, and Pritchard wanted to design a game that fell somewhere between tennis and badminton; they also wanted to make it a game that was fun for children and adults, that emphasized quick reflexes, control, and agility over power and speed, that could be played by singles or doubles, and that was easy to learn. Using secondhand badminton equipment and a weed-infested tennis court, they created, over the course of four or five days, rules drawing from both tennis and badminton, as well as the game's "official" equipment: wooden paddles slightly larger than ping-pong paddles, a plastic whiffle ball, and a three-foot net. (The game is played on a hard surface, and the relatively small size of the court — 44 by 20 feet, about half the size of a tennis court — makes it easy to play in small areas such as driveways or gyms.)

The game quickly found favor with neighbors and friends (Slade Gorton, now a U.S. senator from Washington, had a court built at each of his homes), and soon it spread elsewhere. The majority of serious tournament-quality activity still takes place in the Northwest; the USA Pickle-Ball Association is headquartered in Tacoma, and an estimated 5,000 to 7,000 Puget Sounders play competitively on a regular basis. But the sport has also caught on across the country, especially with junior and senior high school athletic programs; Pickle-Ball Inc., a company formed in Seattle in 1972 by the three inventors, sells balls, paddles, and complete sets to players and groups in all 50 states, and the company's general manager, Doug Smith, estimates the number of recreational players nationwide to be in the hundreds of thousands.

Pictionary

1985

The game of charades, in which teams or individuals silently "act out" a word or phrase and other players guess the meaning, has been a favorite pastime for centuries. In his book *Charades*, James Charlton writes that it may have been invented in France in the 18th century, though an Italian origin is also possible; the Italian word *schiarire* means "to unravel or clarify." Another possible source for the game's name is the word *charrada*, which in Spanish means "the speech of a clown" and in Provençal refers to "chatter." The first mention of a formalized game called charades occurs in a 1776 book called *Mrs. Delaney's Letters*. Sheridan, Sir Walter Scott, and other writers of that era also make reference to it, but the form of charades played then was more of a literary riddle or puzzle. It wasn't until the middle of the 19th century that a complex "acting-out" version became popular, and not until the 1930s — when it was allegedly popularized by Noël Coward and George S. Kaufman — do we find modern charades emerging.

Somewhere during the game's evolution, an anonymous player devised a variation that entailed creating clues by drawing sketches on paper, rather than acting out words in pantomime. In 1985, a Seattle waiter named Rob Angel and two colleagues formalized a set of rules for paper charades, selected a list of words, trademarked a catchy name, and sent the results to the marketplace. They called the game Pictionary.

Angel — who is the first to admit he didn't invent the game, only popularize it — was born in Vancouver, British Columbia, moved at age five to Spokane, and has lived in Seattle since 1984. His exposure to paper charades came after college, when he worked as a waiter in Spokane and shared a house with three other waiters. "It got to be quite a thing. Our girlfriends would show up, and there'd be eight or ten people playing until three or four A.M." Angel toyed with the concept of formalizing the game for three years; in the meantime, he moved to Seattle and began waiting tables at the Lake Union Cafe. In the spring of 1984, Angel decided: "I might as well give it a shot. When the weather got nice, I took a dictionary out into the backyard and started reading. If the word conjured up an image, I'd write it down on a yellow legal pad."

1

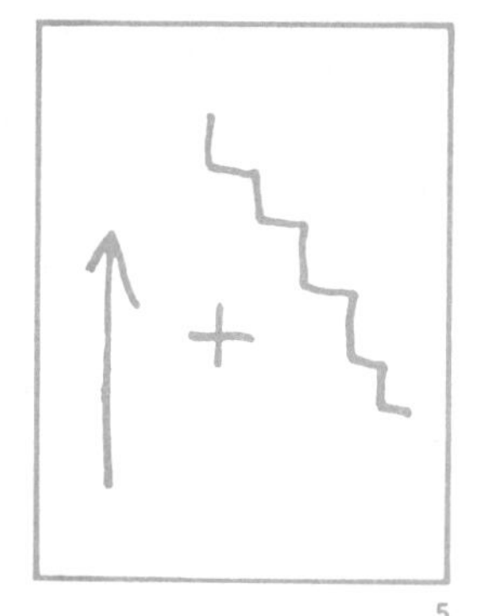

5

Eight months later he had 6,000 words, from which he fine-tuned a final list of 2,500. This list became a key ingredient in the game's success, since some words are more fun to draw and guess than others. He then formalized a simple set of game rules: one member of a team (the "picturist") draws a card containing five words or phrases in categories such as "Person/Place/Animal," "Action," or "Difficult." The picturist then has one minute in which to sketch clues to help teammates guess the word or phrase. If a team succeeds, it gets to roll the dice, move a marker ahead on the scoreboard, and play again. If it fails, it is the other team's turn.

Angel enlisted the aid of Gary Everson, a fellow Lake Union Cafe waiter who had trained as a graphic artist, and Terry Langston, a financial expert whom Angel calls "the money guy," to work out the details of board design and marketing strategy. They registered the word "Pictionary," although the game itself was not patentable, and launched Pictionary Inc. in 1985 with $35,000 borrowed from Angel's uncle. Its popularity was immediate but local at first, and by the end of 1985 the partners had sold some 7,000 games. But it was noticed by Tom McGuire, then a salesman for Selchow & Righter, the firm that launched Trivial Pursuit. In the summer of 1986, McGuire and Angel struck a deal; McGuire quit his job and, along with a few other toy-industry veterans, formed a company called The Games Gang to distribute Pictionary nationally. It became the best-selling board game since Trivial Pursuit, with over 20 million sold worldwide. In addition to the regular version, several variations on the basic theme have been introduced, including a party edition with a dry-erase easel, and a computerized version used in conjunction with Nintendo. Angel is still the president of Pictionary Inc., headquartered in Seattle, and original partners Everson and Langston continue to work with the company.

ANSWER: (1) CRY, (2) ENGAGEMENT RING, (3) SWORDFISH, (4) DIG, (5) UPSTAIRS, (6) MOONLIGHT, (7) REFLECTION, (8) PARTY

Hair Tonic

1940s–1990s

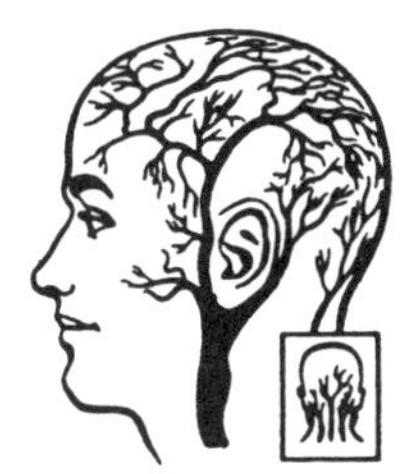

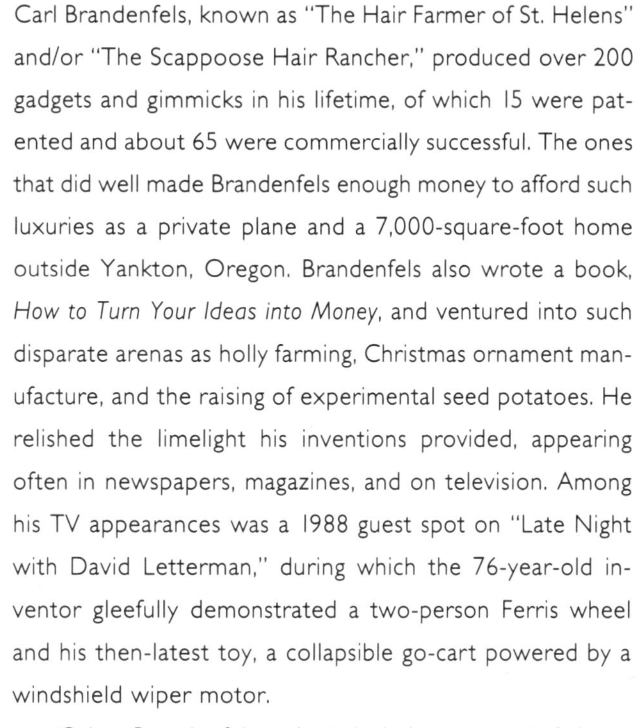

Carl Brandenfels, known as "The Hair Farmer of St. Helens" and/or "The Scappoose Hair Rancher," produced over 200 gadgets and gimmicks in his lifetime, of which 15 were patented and about 65 were commercially successful. The ones that did well made Brandenfels enough money to afford such luxuries as a private plane and a 7,000-square-foot home outside Yankton, Oregon. Brandenfels also wrote a book, *How to Turn Your Ideas into Money*, and ventured into such disparate arenas as holly farming, Christmas ornament manufacture, and the raising of experimental seed potatoes. He relished the limelight his inventions provided, appearing often in newspapers, magazines, and on television. Among his TV appearances was a 1988 guest spot on "Late Night with David Letterman," during which the 76-year-old inventor gleefully demonstrated a two-person Ferris wheel and his then-latest toy, a collapsible go-cart powered by a windshield wiper motor.

Other Brandenfels gadgets include a carpenter's hammer with a three-way level built into the handle; alder cutting boards and butcher blocks; chemically treated pinecones that emitted colors when burned; fruit de-stemmers; sizing rings welded together for easy sorting of garlic harvests; filbert huskers; long-fingered clam diggers; disposable spice grinders; a portable steam engine that burned lawn clippings; and fish bait that exuded synthetic fish oil. One notable exception to his string of successful inventions was "Chef Carl's Old Fashioned German Butter and Egg Fluffy Dumplings," a frozen food that was taken out of production soon after its introduction, when it was discovered that a mysterious

enzyme turned it bright green after six months in the freezer. Brandenfels was stuck with about 10,000 unsold packages of harmless but aesthetically displeasing dumplings.

Brandenfels, who moved to Oregon in 1941, had begun inventing and selling at an early age. As an 11-year-old farm boy in Nebraska, he made a solar shower in 1922 for his family by stretching 400 feet of water-filled garden hose on a barn roof to heat in the sun. He also built crystal sets capable of picking up stations 150 miles away and sold them for five cents apiece.

When Brandenfels began losing his hair in 1936, he created "Brandenfels Scalp and Hair Application," a two-bottle set of sulfanilamide-based tonic with a scalp massager. Brandenfels once estimated that it generated $10 million in sales, a figure especially remarkable considering that Brandenfels himself was bald as an egg. Interviewed in *The Oregonian* in 1984, he remarked, "Well, look at it this way — I'm seventy-three and something has to die. I'm glad it's only my hair." He claimed to have been contacted about the tonic by such celebrities as Frank Sinatra, Ernie Kovacs, Jimmy Durante, and even "a few maharajahs."

One notable exception to the Scappoose Hair Rancher's string of successes: "Chef Carl's Old Fashioned German Butter and Egg Fluffy Dumplings." These were taken out of production when it was discovered that a mysterious enzyme turned them bright green in the freezer. Brandenfels was stuck with about 10,000 packages of harmless but yucky-looking dumplings.

Many of Brandenfels's items were, strictly speaking, not inventions at all. One successful mid-'70s product, created in conjunction with Mary Spitzer of Puget Island, Washington, near Cathlamet, was the "Box of Love," a pretty box with a sentimental poem inside, which was inspired by a gift Spitzer made her daughter one Christmas. Brandenfels and Spitzer sold about 50,000 Boxes of Love at five dollars apiece. In the late 1970s, Brandenfels and Spitzer collaborated again on a system of heat-treating photographs that transferred the pictures' emulsion to clothing and metal. The idea was marketed, with some success, as a clothing novelty and as an aid to retrieving luggage by personalizing it.

List of Illustrations

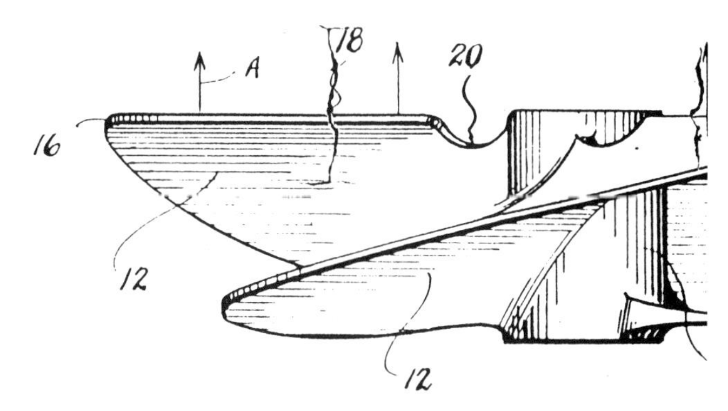

Inventions

Inventors

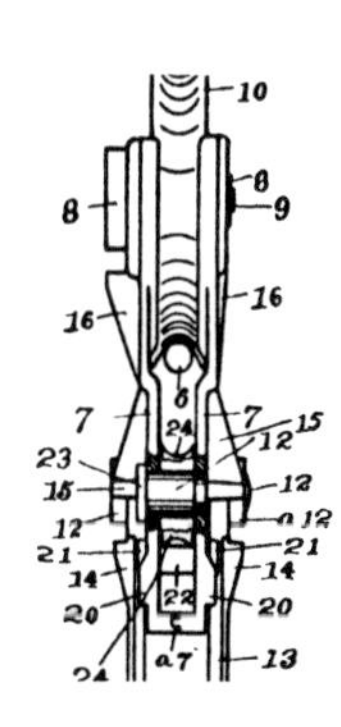

Acknowledgments

I would like to salute — with my self-tipping hat, of course — a few of the people who made writing this book such a pleasure.

It goes without saying that I am indebted to everyone who consented to be interviewed, as well as to the relatives, colleagues, and friends of all the inventors, living and dead, who are represented in this book. Dozens of people along the way had suggestions, as well as juicy rumors for me to pursue. I'd especially like to thank Murray and Rosa Morgan; Archie Satterfield; Don Duncan, Hill Williams, and Rick Anderson of *The Seattle Times*; Hu Blonk of the *Wenatchee World*; Susan English of the *Spokane Spokesman-Review and Chronicle*; Terry Richard and Beth Erickson of *The Oregonian;* David Buerge; J. Kingston Pierce; Dr. James Warren; Grant Fjermedal; Dick McDonald; Esther Mumford; Mildred Andrews; Dan Vorhis; Robert Beach; Richard Seed; Paul and Cecile Andrews; Mary Randlett; Buster Simpson; Jeff Collum; Mike Nelson; the Rand siblings; and Pam Horino of Seal Press.

I am also grateful to the many museum, university, and library staffers who helped, particularly Pam Yorks of the University of Washington Engineering Library Patent Depository; Carla Rickerson and Richard Engeman of the University of Washington Library Special Collections Room; Dr. Paul Spitzer of the Boeing Archives; everyone at the Downtown and Green Lake branches of the Seattle Public Library; Eleanor Toews of the Seattle Public Schools archives; Leila Gray and Kay Rodriguez of the University of Washington Health Sciences Office of News and Information; Bob Roseth of the University of Washington Office of News and Information; Dave Stauth and Andy Duncan of the Oregon State University Office of Communications; Nancy Compau of the Spokane Public Library; Larry Schoonover of the Cheney Cowles Museum in Spokane; Sheryl Stiefel, Martha Fulton, and Carolyn Marr of the Museum of History and Industry in Seattle; Jim Barmore of the Museum of History and Industry and the Skagit County Historical Museum; Sieglinde Smith and Ken Lomax of the Oregon Historical Society Library; Jeff Pedersen of the General Petroleum Museum in Seattle; Bob McCoy of the Museum of Questionable Medical Devices in Minneapolis, Minnesota; and Cheryl Oakes of the Forest Historical Society in Durham, North Carolina.

Early in the game, Stephen Manes, Ed Marquand, and Phyllis Hatfield offered encouragement and valuable advice. Late in the game, Polly Argo and Beth and John Zimmerman provided logistical support. Among others, Stephen Manes, Dr. Paul Spitzer, and Dr. Sandra Downing, by reviewing sections in their areas of expertise, saved me from many embarrassing mistakes. I hasten to add that any and all remaining errors of fact or interpretation are strictly my responsibility.

My family deserves both general and specific thanks. My mother, Ronnie Baumgarten, taught me early on the pleasures of being a bookworm; my father, Alan Woog, demonstrated the value of curiosity. I was lucky enough to have my Old Dad and his friend Ralph Turman as unpaid research assistants for this project; it is a far richer book because of their enthusiasm and energy. (They agreed to help out as long as it didn't cut into the skiing season too much.)

My daughter, Leah, was born while I was writing this, and I am grateful that she put up so good-naturedly with her father, a first-time Mr. Mom. She rarely complained, even when dragged along on interviews or subjected to hours of confinement in my office. And for years now I have been in a state of slightly dazed wonder over the fact that my wife, Karen Kent, lets me get away with being a freelance writer. She is supportive even when the stresses in her own life and work are severe, and she deserves all my thanks and love.

Finally, a word of grateful thanks to the team assembled by Sasquatch Books. Anne Depue, Sasquatch's managing editor, had the idea of collecting Northwest inventors in the first place. She graciously cut me in on the deal and saw the project through with intelligence and good humor. It was both a privilege and great fun to work with Sasquatch's publisher, Chad Haight, and marketing specialists Kitty Harmon and Lynn Tchobanoglous. Designer Scott Hudson of Marquand Books created the visuals. Line editor Phyllis Hatfield and proofreader Barry Foy cleaned up the wayward text. Polly Kenefick and Emily Hall helped find graphics. I could not have asked for better support.

— *A.W.*